I0484289

FEDERAL EXECUTIVE TEAM

Director, Climate Change Science Program: ....William J. Brennan

Director, Climate Change Science Program Office: ....Peter A. Schultz

Lead Agency Principal Representative to CCSP,
Associate Director for Research, Earth Science Division,
National Aeronautics and Space Administration: ....Jack Kaye

Lead Agency Point of Contact, Earth Science Division,
National Aeronautics and Space Administration: ....Hal Maring

Product Lead, Laboratory for Atmospheres,
Earth Science Division, Goddard Space Flight Center,
National Aeronautics and Space Administration: ....Mian Chin

Chair, Synthesis and Assessment Product Advisory Group
Associate Director, National Center for Environmental
Assessment, U.S. Environmental Protection Agency: ....Michael W. Slimak

Synthesis and Assessment Product Coordinator,
Climate Change Science Program Office: ....Fabien J.G. Laurier

EDITORIAL AND PRODUCTION TEAM

Editors: ....Mian Chin, NASA
....Ralph A. Kahn, NASA
....Stephen E. Schwartz, DOE

Graphic Design: ....Sally Bensusen, NASA
....Debbi McLean, NASA

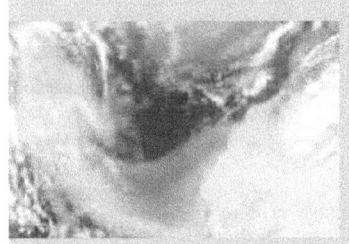

Atmospheric Aerosol Properties and Climate Impacts

Synthesis and Assessment Product 2.3
Report by the U.S. Climate Change Science Program
and the Subcommittee on Global Change Research

COORDINATING LEAD AUTHOR:
Mian Chin, NASA Goddard Space Flight Center

LEAD AND CONTRIBUTING AUTHORS:
Ralph A. Kahn, Lorraine A. Remer, Hongbin Yu, NASA GSFC;
David Rind, NASA GISS;
Graham Feingold, NOAA ESRL; Patricia K. Quinn, NOAA PMEL;
Stephen E. Schwartz, DOE BNL; David G. Streets, DOE ANL;
Philip DeCola, Rangasayi Halthore, NASA HQ

January 2009,

Members of Congress:

On behalf of the National Science and Technology Council, the U.S. Climate Change Science Program (CCSP) is pleased to transmit to the President and the Congress this Synthesis and Assessment Product (SAP) *Atmospheric Aerosol Properties and Climate Impacts*. This is part of a series of 21 SAPs produced by the CCSP aimed at providing current assessments of climate change science to inform public debate, policy, and operational decisions. These reports are also intended to help the CCSP develop future program research priorities.

The CCSP's guiding vision is to provide the Nation and the global community with the science-based knowledge needed to manage the risks and capture the opportunities associated with climate and related environmental changes. The SAPs are important steps toward achieving that vision and help to translate the CCSP's extensive observational and research database into informational tools that directly address key questions being asked of the research community.

This SAP reviews current knowledge about global distributions and properties of atmospheric aerosols, as they relate to aerosol impacts on climate. It was developed in accordance with the Guidelines for Producing CCSP SAPs, the Information Quality Act (Section 515 of the Treasury and General Government Appropriations Act for Fiscal Year 2001 (Public Law 106-554)), and the guidelines issued by the National Aeronautics and Space Administration pursuant to Section 515.

We commend the report's authors for both the thorough nature of their work and their adherence to an inclusive review process.

Sincerely,

Carlos M. Gutierrez
Secretary of Commerce
Chair, Committee on Climate Change
Science and Technology Integration

Samuel W. Bodman
Secretary of Energy
Vice Chair, Committee on Climate
Change Science and Technology
Integration

John H. Marburger III
Director, Office of Science and
Technology Policy
Executive Director, Committee
on Climate Change Science and
Technology Integration

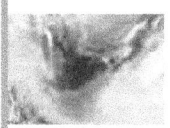

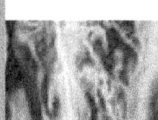

Executive Summary	Lorraine A. Remer, NASA GSFC; Mian Chin, NASA GSFC; Philip DeCola, NASA HQ; Graham Feingold, NOAA ERSL; Rangasayi Halthore, NASA HQ/NRL; Ralph A. Kahn, NASA GSFC; Patricia K. Quinn, NOAA PMEL; David Rind, NASA GISS; Stephen E. Schwartz, DOE BNL; David G. Streets, DOE ANL; Hongbin Yu, NASA GSFC/UMBC
Chapter 1	**Lead Authors:** Ralph A. Kahn, NASA GSFC; Hongbin Yu, NASA GSFC/UMBC **Contributing Authors:** Stephen E. Schwartz, DOE BNL; Mian Chin, NASA GSFC; Graham Feingold, NOAA ESRL; Lorraine A. Remer, NASA GSFC; David Rind, NASA GISS; Rangasayi Halthore, NASA HQ/NRL; Philip DeCola, NASA HQ
Chapter 2	**Lead Authors:** Hongbin Yu, NASA GSFC/UMBC; Patricia K. Quinn, NOAA PMEL; Graham Feingold, NOAA ESRL; Lorraine A. Remer, NASA GSFC; Ralph A. Kahn, NASA GSFC **Contributing Authors:** Mian Chin, NASA GSFC; Stephen E. Schwartz, DOE BNL
Chapter 3	**Lead Authors:** David Rind, NASA GISS; Mian Chin, NASA GSFC; Graham Feingold, NOAA ESRL; David G. Streets, DOE ANL **Contributing Authors:** Ralph A. Kahn, NASA GSFC; Stephen E. Schwartz, DOE BNL; Hongbin Yu, NASA GSFC/UMBC
Chapter 4	David Rind, NASA GISS; Ralph A. Kahn, NASA GSFC; Mian Chin, NASA GSFC; Stephen E. Schwartz, DOE BNL; Lorraine A. Remer, NASA GSFC; Graham Feingold, NOAA ESRL; Hongbin Yu, NASA GSFC/UMBC; Patricia K. Quinn, NOAA PMEL; Rangasayi Halthore, NASA HQ/NRL

ACKNOWLEDGMENTS

First, the authors wish to acknowledge the late Yoram J. Kaufman both for his inspiration and contributions to aerosol-climate science throughout his career and for his early leadership of the activity that produced this document. His untimely passing left it to the remaining authors to complete this report. Yoram and his contributions to our community are greatly missed.

This Climate Change Science Program Synthesis and Assessment Product (CCSP SAP) 2.3 has been reviewed by a group of experts, the public, and Federal Agencies. The purpose of these independent reviews was to assure the quality of this product.

We wish to thank the following individuals for their expert review of this report: Sundar Christopher (University of Alabama Huntsville), Daniel Jacob (Harvard University), Steven Ghan (Pacific Northwest National Laboratory), John Ogren (NOAA Earth System Research Laboratory), and Susan Solomon (NOAA Earth System Research Laboratory).

We also wish to thank the following individuals/group for their public/federal agency review of this report: Joel D. Scheraga (EPA), Samuel P. Williamson (NOAA/OFCM), Alan Carlin, David L. Hagen, Douglas Hoyt, Forrest M. Mims III (Geronimo Creek observatory), John Pittman, Nathan Taylor (Texas A&M University), Werner Weber (Technische University Dortmund, Germany), and the NOAA Research Council.

The work by Bates et al. (2006), Penner et al. (2006), Yu et al. (2006), Textor et al. (2006), Kinne et al. (2006), Schulz et al. (2006), and the Fourth Assessment Report of the Intergovernmental Panel on Climate Change (2007) provided important groundwork for the material in Chapter 2 and Chapter 3.

RECOMMENDED CITATIONS

For the Report as a Whole:

CCSP 2009: *Atmospheric Aerosol Properties and Climate Impacts,* A Report by the U.S. Climate Change Science Program and the Subcommittee on Global Change Research. [Mian Chin, Ralph A. Kahn, and Stephen E. Schwartz (eds.)]. National Aeronautics and Space Administration, Washington, D.C., USA, 128 pp.

For the Executive Summary:

Remer, L. A., M. Chin, P. DeCola, G. Feingold, R. Halthore, R. A. Kahn, P. K. Quinn, D. Rind, S. E. Schwartz, D. Streets, and H. Yu, 2009: Executive Summary, in *Atmospheric Aerosol Properties and Climate Impacts,* A Report by the U.S. Climate Change Science Program and the Subcommittee on Global Change Research. [Mian Chin, Ralph A. Kahn, and Stephen E. Schwartz (eds.)]. National Aeronautics and Space Administration, Washington, D.C., USA.

For Chapter 1:

Kahn, R. A., H. Yu, S. E. Schwartz, M. Chin, G. Feingold, L. A. Remer, D. Rind, R. Halthore, and P. DeCola, 2009: Introduction, in *Atmospheric Aerosol Properties and Climate Impacts,* A Report by the U.S. Climate Change Science Program and the Subcommittee on Global Change Research. [Mian Chin, Ralph A. Kahn, and Stephen E. Schwartz (eds.)]. National Aeronautics and Space Administration, Washington, D.C., USA.

For Chapter 2:

Yu, H., P. K. Quinn, G. Feingold, L. A. Remer, R. A. Kahn, M. Chin, and S. E. Schwartz, 2009: Remote Sensing and *In Situ* Measurements of Aerosol Properties, Burdens, and Radiative Forcing, in *Atmospheric Aerosol Properties and Climate Impacts,* A Report by the U.S. Climate Change Science Program and the Subcommittee on Global Change Research. [Mian Chin, Ralph A. Kahn, and Stephen E. Schwartz (eds.)]. National Aeronautics and Space Administration, Washington, D.C., USA.

For Chapter 3:

Rind, D., M. Chin, G. Feingold, D. Streets, R. A. Kahn, S. E. Schwartz, and H. Yu, 2009: Modeling the Effects of Aerosols on Climate, in *Atmospheric Aerosol Properties and Climate Impacts,* A Report by the U.S. Climate Change Science Program and the Subcommittee on Global Change Research. [Mian Chin, Ralph A. Kahn, and Stephen E. Schwartz (eds.)]. National Aeronautics and Space Administration, Washington, D.C., USA.

For Chapter 4:

Rind, D., R. A. Kahn, M. Chin, S. E. Schwartz, L. A. Remer, G. Feingold, H. Yu, P. K. Quinn, and R. Halthore, 2009: The Way Forward, in *Atmospheric Aerosol Properties and Climate Impacts,* A Report by the U.S. Climate Change Science Program and the Subcommittee on Global Change Research. [Mian Chin, Ralph A. Kahn, and Stephen E. Schwartz (eds.)]. National Aeronautics and Space Administration, Washington, D.C., USA.

Earth observed from space. Much of the information contained in this image came from the MODIS instrument on the NASA Terra satellite. This 2002 "Blue Marble" features land surfaces, clouds, topography, and city lights. Credit: NASA (image processed by Robert Simmon and Reto Stöckli).

Authors: Lorraine A. Remer, NASA GSFC; Mian Chin, NASA GSFC; Philip DeCola, NASA HQ; Graham Feingold, NOAA ERSL; Rangasayi Halthore, NASA HQ/NRL; Ralph A. Kahn, NASA GSFC; Patricia K. Quinn, NOAA PMEL; David Rind, NASA GISS; Stephen E. Schwartz, DOE BNL; David G. Streets, DOE ANL; Hongbin Yu, NASA GSFC/UMBC

This report critically reviews current knowledge about global distributions and properties of atmospheric aerosols, as they relate to aerosol impacts on climate. It assesses possible next steps aimed at substantially reducing uncertainties in aerosol radiative forcing estimates. Current measurement techniques and modeling approaches are summarized, providing context. As a part of the Synthesis and Assessment Product in the Climate Change Science Program, this assessment builds upon recent related assessments, including the Fourth Assessment Report of the Intergovernmental Panel on Climate Change (IPCC AR4, 2007) and other Climate Change Science Program reports. The objectives of this report are (1) to promote a consensus about the knowledge base for climate change decision support, and (2) to provide a synthesis and integration of the current knowledge of the climate-relevant impacts of anthropogenic aerosols for policy makers, policy analysts, and general public, both within and outside the U.S government and worldwide.

ES 1. AEROSOLS AND THEIR CLIMATE EFFECTS

ES 1.1. Atmospheric Aerosols

Atmospheric aerosols are suspensions of solid and/or liquid particles in air. Aerosols are ubiquitous in air and are often observable as dust, smoke, and haze. Both natural and human processes contribute to aerosol concentrations. On a global basis, aerosol mass derives predominantly from natural sources, mainly sea salt and dust. However, anthropogenic (manmade) aerosols, arising primarily from a variety of combustion sources, can dominate in and downwind of highly populated and industrialized regions, and in areas of intense agricultural burning.

The term "atmospheric aerosol" encompasses a wide range of particle types having different compositions, sizes, shapes, and optical properties. Aerosol loading, or amount in the atmosphere, is usually quantified by mass concentration or by an optical measure, aerosol optical depth (AOD). AOD is the vertical integral through the entire height of the atmosphere of the fraction of incident light either scattered or absorbed by airborne particles. Usually numerical models and *in situ* observations use mass concentration as the primary measure of aerosol loading, whereas most remote sensing methods retrieve AOD.

ES 1.2. Radiative Forcing of Aerosols

Aerosols affect Earth's energy budget by scattering and absorbing radiation (the "direct effect") and by modifying amounts and microphysical and radiative properties of clouds (the "indirect effects"). Aerosols influence cloud properties through their role as cloud condensation nuclei (CCN) and/or ice nuclei. Increases in aerosol particle concentrations may increase the ambient concentration of CCN and ice nuclei, affecting cloud properties. A CCN increase can lead to more cloud droplets so that, for fixed cloud liquid water content, the cloud droplet size will decrease. This effect leads to brighter clouds (the "cloud albedo effect"). Aerosols can also affect clouds by absorbing solar energy and altering the environment in which the cloud develops, thus changing cloud properties without actually serving as CCN. Such effects can change precipitation patterns as well as cloud extent and optical properties.

The addition of aerosols to the atmosphere alters the intensity of sunlight scattered back to space, absorbed in the atmosphere, and arriving

Aerosols affect Earth's energy budget by scattering and absorbing radiation (the "direct effect") and by modifying amounts and microphysical and radiative properties of clouds (the "indirect effects").

at the surface. Such a perturbation of sunlight by aerosols is designated aerosol radiative forcing (RF). Note that RF must be defined as a perturbation from an initial state, whether that state be the complete absence of aerosols, the estimate of aerosol loading from pre-industrial times, or an estimate of aerosol loading for today's natural aerosols. The RF calculated from the difference between today's total aerosol loading (natural plus anthropogenic) and each of the three initial states mentioned above will result in different values. Also, the aerosol RF calculated at the top of the atmosphere, the bottom of the atmosphere, or any altitude in between, will result in different values. Other quantities that need to be specified when reporting aerosol RF include the wavelength range, the temporal averaging, the cloud conditions considered for direct effects, and the aerosol-cloud interactions that are being considered for the broad classifications of indirect and semi-direct effects. Regardless of the exact definition of aerosol RF, it is characterized by large spatial and temporal heterogeneity due to the wide variety of aerosol sources and types, the spatial non-uniformity and intermittency of these sources, the short atmospheric lifetime of aerosols, and the chemical and microphysical processing that occurs in the atmosphere.

On a global average basis, the sum of direct and indirect forcing by anthropogenic aerosols at the top of the atmosphere is almost certainly negative (a cooling influence), and thus almost certainly offsets a fraction of the positive (warming) forcing due to anthropogenic greenhouse gases. However, because of the spatial and temporal non-uniformity of the aerosol RF, and likely differences in the effects of shortwave and longwave forcings, the net effect on Earth's climate is not simply a fractional offset to the effects of forcing by anthropogenic greenhouse gases.

ES 1.3. Reducing Uncertainties in Aerosol Radiative Forcing Estimates

The need to represent aerosol influences on climate is rooted in the larger, policy related requirement to predict the climate changes that would result from different future emission strategies. This requires that confidence in climate models be based on their ability to accurately represent not just present climate, but also the changes that have occurred over

roughly the past century. Achieving such confidence depends upon adequately understanding the forcings that have occurred over this period. Although the forcing by long-lived greenhouse gases is known relatively accurately for this period, the history of total forcing is not, due mainly to the uncertain contribution of aerosols.

Present-day aerosol radiative forcing relative to preindustrial is estimated primarily using numerical models that simulate the emissions of aerosol particles and gaseous precursors and the aerosol and cloud processes in the atmosphere. The accuracy of the models is assessed primarily by comparison with observations. The key to reducing aerosol RF uncertainty estimates is to understand the contributing processes well enough to accurately reproduce them in models. This report assesses present ability to represent in models the distribution, properties and forcings of present-day aerosols, and examines the limitations of currently available models and measurements. The report identifies three specific areas where continued, focused effort would likely result in substantial reduction in present-day aerosol forcing uncertainty estimates: (1) improving quality and coverage of aerosol measurements, (2) achieving more effective use of these measurements to constrain model simulation/assimilation and to test model parameterizations, and (3) producing more accurate representation of aerosols and clouds in models.

ES 2. MEASUREMENT-BASED ASSESSMENT OF AEROSOL RADIATIVE FORCING

Over the past decade, measurements of aerosol amount, geographical distribution, and physical and chemical properties have substantially improved, and understanding of the controlling processes and the direct and indirect radiative effects of aerosols has increased. Key research activities have been:

*   Development and implementation of new and enhanced satellite-borne sensors capable of observing the spatial and temporal characteristics of aerosol properties and examine aerosol effects on atmospheric radiation.
*   Execution of focused field experiments examining aerosol processes and properties in various aerosol regimes around the globe;

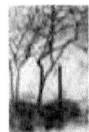

Forcing by anthropogenic aerosols at the top of the atmosphere is negative (cooling) and offsets a fraction of the positive (warming) forcing by greenhouse gases. However, because of the spatial and temporal non-uniformity of aerosol forcing, the net effect is not simply a fractional offset.

- Establishment and enhancement of ground-based networks measuring aerosol properties and radiative effects;
- Development and deployment of new and enhanced instrumentation including devices to determine size dependent particle composition on fast timescales, and methods for determining aerosol light absorption coefficients and single scattering albedo.

ES 2.1. Assessments of Aerosol Direct Radiative Forcing

Over the past 15 years, focused field campaigns have provided detailed characterizations of regional aerosol, chemical, microphysical and radiative properties, along with relevant surface and atmospheric conditions. Studies from these campaigns provide highly reliable characterization of submicrometer spherical particles such as sulfate and carbonaceous aerosol. *In situ* characterization of larger particles such as dust are much less reliable.

For all their advantages, field campaigns are inherently limited by their relatively short duration and small spatial coverage. Surface networks and satellites provide a needed long-term view, and satellites provide additional extensive spatial coverage. Surface networks, such as the Aerosol Robotic Network (AERONET), provide observations of AOD at mid-visible wavelengths with an accuracy of 0.01 to 0.02, nearly three to five times more accurate than satellite retrievals. These same remote sensing ground networks also typically retrieve column integrated aerosol microphysical properties, but with uncertainties that are much larger than *in situ* measurements.

The satellite remote sensing capability developed over the past decades has enabled the estimate of aerosol radiative forcing on a global scale. Current satellite sensors such as the MODerate resolution Imaging Spectroradiometer (MODIS) and Multi-angle Imaging SpectroRadiometer (MISR) can retrieve AOD (τ) under cloud free conditions with an accuracy of $\pm0.05 \pm 0.20\tau$ over land and better than $\pm0.04 \pm 0.1\tau$ over ocean at mid-visible wavelength. In addition, these and other satellite sensors can qualitatively retrieve particle properties (size, shape and absorption), a major advance over the previous generation of satellite instruments. Much effort has gone into comparing different observational methods to

estimate global oceanic cloud-free aerosol direct radiative forcing for solar wavelengths at the top of the atmosphere (TOA). Applying various methods using MODIS, MISR and the Clouds and Earth's Radiant Energy System (CERES), the aerosol direct RF at TOA derived above ocean converges to -5.5 ± 0.2 W m^{-2}, where the initial state of the forcing perturbation is a completely aerosol-free atmosphere. Here, the uncertainty is the standard deviation of the various methods, indicating close agreement between the different satellite data sets. However, regional comparisons of the various methods show greater spread than the global mean. Estimates of direct radiative forcing at the ocean surface, and at top and bottom of the atmosphere over land, are also reported, but are much less certain. All these measurement-based estimates are calculated for cloud-free conditions using an initial state of an aerosol-free atmosphere.

Although no proven methods exist for measuring the anthropogenic component of the observed aerosol over broad geographic regions, satellite retrievals are able to qualitatively determine aerosol type under some conditions. From observations of aerosol type, the best estimates indicate that approximately 20% of the AOD over the global oceans is a result of human activities. Following from these estimates of anthropogenic fraction, the cloud-free anthropogenic direct radiative forcing at TOA is approximated to be -1.1 ± 0.4 W m^{-2} over the global ocean, representing the anthropogenic perturbation to today's natural aerosol.

ES 2.2. Assessments of Aerosol Indirect Radiative Forcing

Remote sensing estimates of aerosol indirect forcing are still very uncertain. Even on small spatial scales, remote sensing of aerosol effects on cloud albedo do not match *in situ* observations, due to a variety of difficulties with the remote sensing of cloud properties at fine scales, the inability of satellites to observe aerosol properties beneath cloud base, and the difficulty of making aerosol retrievals in cloud fields. Key quantities such as liquid water path, cloud updraft velocity and detailed aerosol size distributions are rarely constrained by coincident observations.

Most remote sensing observations of aerosol-cloud interactions and aerosol indirect forcing are based on simple correlations among variables, which do not establish cause-and-effect relationships. Inferring aerosol effects on clouds from the observed relationships is complicated further because aerosol loading and meteorology are often correlated, making it difficult to distinguish aerosol from meteorological effects. As in the case of direct forcing, the regional nature of indirect forcing is especially important for understanding actual climate impact.

ES 3. MODEL ESTIMATED AEROSOL RADIATIVE FORCING AND ITS CLIMATE IMPACT

Just as different types of aerosol observations serve similar purposes, diverse types of models provide a variety of approaches to understanding aerosol forcing of climate. Large-scale Chemistry and Transport Models (CTMs) are used to test current understanding of the processes controlling aerosol spatial and temporal distributions, including aerosol and precursor emissions, chemical and microphysical transformations, transport, and removal. CTMs are used to describe the global aerosol system and to make estimates of direct aerosol radiative forcing. In general, CTMs do not explore the climate response to this forcing. General Circulation Models (GCMs), sometimes called Global Climate Models, have the capability of including aerosol processes as a part of the climate system to estimate aerosol climate forcing, including aerosol-cloud interactions, and the climate response to this forcing. Another type of model represents atmospheric processes on much smaller scales, such as cloud resolving and large eddy simulation models. These small-scale models are the primary tools for improving understanding of aerosol-cloud processes, although they are not used to make estimates of aerosol-cloud radiative forcing on regional or global scales.

ES 3.1. The Importance of Aerosol Radiative Forcing in Climate Models
Calculated change of surface temperature due to forcing by anthropogenic greenhouse gases and aerosols was reported in IPCC AR4 based on results from more than 20 participating global climate modeling groups. Despite a wide range of climate sensitivity (i.e. the amount of surface temperature increase due to a change in radiative forcing, such as an increase of CO_2) exhibited by the models, they all yield a global

average temperature change very similar to that observed over the past century. This agreement across models appears to be a consequence of the use of very different aerosol forcing values, which compensates for the range of climate sensitivity. For example, the direct cooling effect of sulfate aerosol varied by a factor of six among the models. An even greater disparity was seen in the model treatment of black carbon and organic carbon. Some models ignored aerosol indirect effects whereas others included large indirect effects. In addition, for those models that included the indirect effect, the aerosol effect on cloud brightness (reflectivity) varied by up to a factor of nine. Therefore, the fact that models have reproduced the global temperature change in the past does not imply that their future forecasts are accurate. This state of affairs will remain until a firmer estimate of radiative forcing by aerosols, as well as climate sensitivity, is available.

ES 3.2. Modeling Atmospheric Aerosols
Simulations of the global aerosol distribution by different models show good agreement in their representation of the global mean AOD, which in general also agrees with satellite-observed values. However, large differences exist in model simulations of regional and seasonal distributions of AOD, and in the proportion of aerosol mass attributed to individual species. Each model uses its own estimates of aerosol and precursor emissions and configurations for chemical transformations, microphysical properties, transport, and deposition. Multi-model experiments indicate that differences in the models' atmospheric processes play a more important role than differences in emissions in creating the diversity among model results. Although aerosol mass concentration is the basic measure of aerosol loading in the models, this quantity is translated to AOD via mass extinction efficiency in order to compare with observations and then to estimate aerosol direct RF. Each model employs its own mass extinction efficiency based on limited knowledge of optical and physical properties of each aerosol type. Thus, it is possible for the models to produce different distributions of aerosol loading as mass concentrations but agree in their distributions of AOD, and vice-versa.

Model calculated total global mean direct anthropogenic aerosol RF at TOA, based on the difference between pre-industrial and current

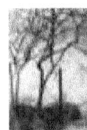

The fact that models have reproduced the global temperature change in the past does not imply that their future forecasts are accurate. This state of affairs will remain until a firmer estimate of radiative forcing by aerosols, as well as climate sensitivity, is available.

aerosol fields, is -0.22 W m-2, with a range from -0.63 to +0.04 W m-2. This estimate does not include man-made contributions of nitrate and dust, which could add another -0.2 W m-2 estimated by IPCC AR4. The mean value is much smaller than the estimates of total greenhouse gas forcing of +2.9 W m-2, but the comparison of global average values does not take into account immense regional variability. Over the major sources and their downwind regions, the model-calculated negative forcing from aerosols can be comparable to or even larger than the positive forcing by greenhouse gases.

ES 3.3. Aerosol Effects on Clouds

Large-scale models are increasingly incorporating aerosol indirect effects into their calculations. Published large-scale model studies report calculated global cloud albedo effect RF at top-of-atmosphere, based on the perturbation from pre-industrial aerosol fields, ranging from -0.22 to -1.85 W m-2 with a central value of -0.7 W m-2. Numerical experiments have shown that the cloud albedo effect is not a strong function of a model's cloud or radiation scheme, and that although model representations of cloud physics are important, the differences in modeled aerosol concentrations play a strong role in inducing differences in the indirect as well as the direct effect. Although small-scale models, such as cloud-resolving or large eddy simulation models, do not attempt to estimate global aerosol RF, they are essential for understanding the fundamental processes occurring in clouds, which then leads to better representation of these processes in larger-scale models.

ES 3.4. Impacts of Aerosols on Climate Model Simulations

The current aerosol modeling capability demonstrated by chemical transport models has not been fully incorporated into GCM simulations. Of the 20+ models used in the IPCC AR4 assessment, most included sulfate direct RF, but only a fraction considered other aerosol types, and only less than a third included aerosol indirect effects. The lack of a comprehensive representation of aerosols in climate models makes it difficult to determine climate sensitivity, and thus to make climate change predictions.

Although the nature and geographical distribution of forcings by greenhouse gases and aerosols are quite different, it is often assumed that to first approximation the effects of these

forcings on global mean surface temperature are additive, so that the negative forcing by anthropogenic aerosols has partially offset the positive forcing by incremental greenhouse gas increases over the industrial period. The IPCC AR4 estimates the total global average TOA forcing by incremental greenhouse gases to be 2.9 ± 0.3 W m-2, where the uncertainty range is meant to encompass the 90% probability that the actual value will be within the indicated range. The corresponding value for aerosol forcing at TOA (direct plus enhanced cloud albedo effects), defined as the perturbation from pre-industrial conditions, is -1.3 (-2.2 to -0.5) W m-2. The total forcing, 1.6 (0.6 to 2.4) W m-2, reflects the offset of greenhouse gas forcing by aerosols, where the uncertainty in total anthropogenic RF is dominated by the uncertainty in aerosol RF.

However, since aerosol forcing is much more pronounced on regional scales than on the global scale because of the highly variable aerosol distributions, it would be insufficient or even misleading to place too much emphasis on the global average. Also, aerosol RF at the surface is stronger than that at TOA, exerting large impacts within the atmosphere to alter the atmospheric circulation patterns and water cycle. Therefore, impacts of aerosols on climate should be assessed beyond the limted aspect of globally averaged radiative forcing at TOA.

ES 4. THE WAY FORWARD

The uncertainty in assessing total anthropogenic greenhouse gas and aerosol impacts on climate must be much reduced from its current level to allow meaningful predictions of future climate. This uncertainty is currently dominated by the aerosol component. In addition, evaluation of aerosol effects on climate must take into account high spatial and temporal variation of aerosol amounts and properties as well as the aerosol interactions with clouds and precipitation. Thus, the way forward requires more certain estimates of aerosol radiative forcing, which in turn requires better observations, improved models, and a synergistic approach.

From the observational perspective, the high priority tasks are:
- **Maintain current and enhance future satellite capabilities** for measuring geographical and vertical distribution of aerosol

The uncertainty in total anthropogenic radiative forcing (greenhouse gases + aerosols) is dominated by the uncertainty in aerosol radiative forcing.

Impacts of aerosols on climate should be assessed beyond the limited aspect of globally-averaged radiative forcing at top-of-atmosphere.

amount and optical properties, suitable for estimating aerosol forcing over multi-decadal time scales and for evaluating global models.

- **Maintain, enhance, and expand the surface observation networks** measuring aerosol optical properties for satellite retrieval validation, model evaluation, and climate change assessments. Observation should be augmented with routine measurements of other key parameters with state-of-art techniques.

- **Execute a continuing series of coordinated field campaigns** aiming to study the atmospheric processes, to broaden the database of detailed aerosol chemical, physical, and optical/radiative characteristics, to validate remote-sensing retrieval products, and to evaluate chemistry transport models.

- **Initiate and carry out a systematic program of simultaneous measurement of aerosol** composition and size distribution, cloud microphysical properties, and precipitation variables.

- **Fully exploit the existing information in satellite observations of AOD and particle type** by refining retrieval algorithms, quantifying data quality, extracting greater aerosol information from joint multi-sensor products, and generating uniform, climate-quality data records.

- **Measure the formation, evolution, and properties of aerosols under controlled laboratory conditions** to develop mechanistic and quantitative understanding of aerosol formation, chemistry, and dynamics.

- **Improve measurement-based techniques for distinguishing anthropogenic from natural aerosols** by combining satellite data analysis with *in situ* measurements and modeling methods.

Individual sensors or instruments have both strengths and limitations, and no single strategy is adequate for characterizing the complex aerosol system. The best approach is to make synergistic use of measurements from multiple platforms, sensors and instruments having complementary capabilities. The wealth of information coming from the variety of today's sensors has not yet been fully exploited. Advances in measurement-based estimates of aerosol radiative forcing are expected in the near future, as existing data sets are more fully

explored. Even so, the long-term success in reducing climate-change prediction uncertainties rests with improving modeling capabilities, and today's suite of observations can only go so far towards that goal.

From the modeling perspective, the high priority tasks are:

- **Improve the accuracy and capability of model simulation of aerosols** (including components and atmospheric processes) and aerosol direct radiative forcing. Observational strategies described above must be developed to constrain and validate the key parameters in the model.

- **Advance the ability to model aerosol-cloud-precipitation interaction in climate models, particularly the simulation of clouds,** in order to reduce the largest uncertainty in the climate forcing/feedback processes.

- **Incorporate improved representation of aerosol processes in coupled aerosol-climate system models** and evaluate the ability of these models to simulate present climate and past (twentieth century) climate change.

- **Apply coupled aerosol-climate system models to assess the climate change** that would result from alternative scenarios of prospective future emissions of greenhouse gases and aerosols and aerosol precursors.

In addition to the above priorities in measurements and modeling, there is a critical need to:

- **Develop and evaluate emission inventories of aerosol particles and precursor gases.** Continuous development and improvement of current emissions, better estimates of past emissions, and projection of future emissions should be maintained.

Progress in improving modeling capabilities requires effort on the observational side, to reduce uncertainties and disagreements among observational data sets. The way forward will require integration of satellite and *in situ* measurements into global models. However, understanding the strengths and weaknesses of each observational data set must be clear in order for the constraints they provide to improve confidence in the models, and for efforts at data assimilation to succeed.

The way forward requires more certain estimates of aerosol radiative forcing, which in turn requires better observations, improved models, and a synergistic approach.

Narrowing the gap between the current understanding of long-lived greenhouse gas and that of anthropogenic aerosol contributions to RF will require progress in all aspects of aerosol-climate science. Development of new space-based, field and laboratory instruments will be needed, and in parallel, more realistic simulations of aerosol, cloud and atmospheric processes must be incorporated into models. Most importantly, greater synergy among different types of measurements, among different types of models, and especially between measurements and models is critical. Aerosol-climate science will naturally expand to encompass not only radiative effects on climate, but also aerosol effects on cloud processes, precipitation, and weather. New initiatives will strive to more effectively include experimentalists, remote sensing scientists and modelers as equal partners, and the traditionally defined communities in different atmospheric science disciplines will increasingly find common ground in addressing the challenges ahead.

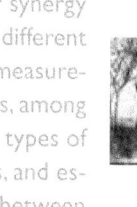

Most importantly, greater synergy among different types of measurements, among different types of models, and especially between measurements and models is critical.

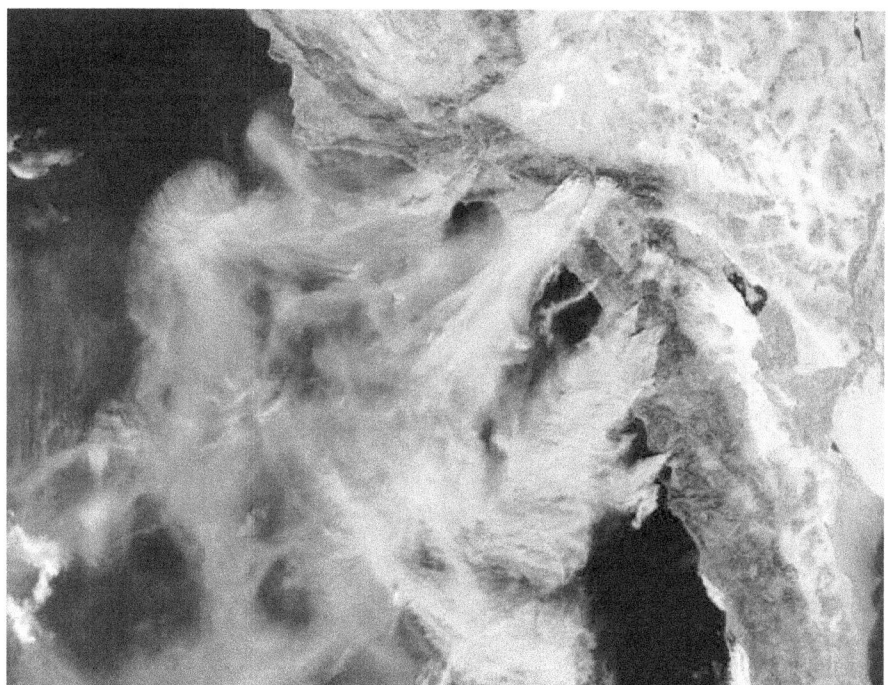

Several massive wildfires were across southern California during October 2003. MODIS, on the NASA Terra satellite, captured smoke spreading across the region and westward over the Pacific Ocean on October 26, 2003. Credit: NASA.

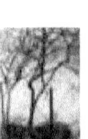

Mexico city, located in a basin surrounded by mountains, often accumulates air pollution—anthropogenic combustion particles, sometimes mixed with wildfire smoke and mineral dust from the surrounding region. Photo taken from the NASA DC-8 aircraft during the INTEX-B field experiment in spring 2006. Credit: Cameron McNaughton, University of Hawaii.

Los Angeles in the haze at sunset. Pollution aerosols scatter sunlight, shrouding the region in an intense orange-brown glow, as seen through an airplane window, looking west across the LA River, with the city skyline in the background. Credit: Barbara Gaitley, JPL/NASA.

Introduction

Lead Authors: Ralph A. Kahn, NASA GSFC; Hongbin Yu, NASA GSFC/UMBC
Contributing Authors: Stephen E. Schwartz, DOE BNL; Mian Chin, NASA GSFC; Graham Feingold, NOAA ESRL; Lorraine A. Remer, NASA GSFC; David Rind, NASA GISS; Rangasayi Halthore, NASA HQ/NRL; Philip DeCola, NASA HQ

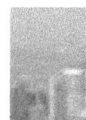

This report highlights key aspects of current knowledge about the global distribution of aerosols and their properties, as they relate to climate change. Leading measurement techniques and modeling approaches are briefly summarized, providing context for an assessment of the next steps needed to significantly reduce uncertainties in this component of the climate change picture. The present assessment builds upon the recent Inter-governmental Panel on Climate Change Fourth Assessment Report (IPCC AR4, 2007) and other sources.

1.1 Description of Atmospheric Aerosols

Although Earth's atmosphere consists primarily of gases, aerosols and clouds play significant roles in shaping conditions at the surface and in the lower atmosphere. Aerosols are liquid or solid particles suspended in the air, whose typical diameters range over four orders of magnitude, from a few nanometers to a few tens of micrometers. They exhibit a wide range of compositions and shapes, that depend on the their origins and subsequent atmospheric processing. For many applications, aerosols from about 0.05 to 10 micrometers in diameter are of greatest interest, as particles in this size range dominate aerosol direct interaction with sunlight, and also make up the majority of the aerosol mass. Particles at the small end of this size range play a significant role in interactions with clouds, whereas particles at the large end, though much less numerous, can contribute significantly near dust and volcanic sources. Over the ocean, giant salt particles may also play a role in cloud development.

A large fraction of aerosols is natural in origin, including desert and soil dust, wildfire smoke, sea salt particles produced mainly by breaking bubbles in the spray of ocean whitecaps, and volcanic ash. Volcanoes are also sources of sulfur dioxide, which, along with sulfur-containing gases produced by ocean biology and the decomposition of organic matter, as well as hydrocarbons such as terpenes and isoprene emitted by vegetation, are examples of gases that can be converted to so-called "secondary" aerosols by chemical processes in the atmosphere. Figure 1.1 gives a summary of aerosol processes most relevant to their influence on climate.

Table 1.1 reports estimated source strengths, lifetimes, and amounts for major aerosol types, based on an aggregate of emissions estimates and global model simulations; the ranges provided represent model diversity only, as the global measurements required to validate these quantities are currently lacking.

Aerosol optical depth (AOD) (also called aerosol optical thickness, AOT, in the literature) is a measure of the amount of incident light either scattered or absorbed by airborne particles. Formally, aerosol optical depth is a dimensionless quantity, the integral of the product of particle number concentration and particle extinction cross-section (which accounts for individual particle scattering + absorption), along a path length through the atmosphere, usually measured vertically. In addition to AOD, particle size, composition, and structure, which are mediated both by source type and subsequent atmospheric processing, determine how particles interact with radiant energy and influence the heat balance of the planet. Size and composition also determine the ability of particles to serve as nuclei upon which cloud droplets form. This provides an indirect means for aerosol to interact with radiant energy by modifying cloud properties.

One of the greatest challenges in studying aerosol impacts on climate is the immense diversity, not only in particle size, composition, and origin, but also in spatial and temporal distribution.

Among the main aerosol properties required to evaluate their effect on radiation is the *single-scattering albedo* (SSA), which describes the fraction of light interacting with the particle that is scattered, compared to the total that is scattered and absorbed. Values range from 0 for totally absorbing (dark) particles to 1 for purely scattering ones; in nature, SSA is rarely lower than about 0.75. Another quantity, the *asymmetry parameter* (g), reports the first moment of the cosine of the scattered radiation angular distribution. The parameter g ranges from -1 for entirely back-scattering particles, to 0 for isotropic (uniform) scattering, to +1 for entirely forward-scattering. One further quantity that

must be considered in the energy balance is the *surface albedo* (A), a measure of reflectivity at the ground, which, like SSA, ranges from 0 for purely absorbing to 1 for purely reflecting. In practice, A can be near 0 for dark surfaces, and can reach values above 0.9 for visible light over snow. AOD, SSA, g, and A are all dimensionless quantities, and are in general wavelength-dependent. In this report, AOD, SSA, and g are given at mid-visible wavelengths, near the peak of the solar spectrum around 550 nanometers, and A is given as an average over the solar spectrum, unless specified otherwise.

About 10% of global atmospheric aerosol mass is generated by human activity, but it is concentrated in the immediate vicinity, and downwind of sources (e.g., Textor et al., 2006). These anthropogenic aerosols include primary (directly emitted) particles and secondary particles that are formed in the atmosphere. Anthropogenic aerosols originate from urban and industrial emissions, domestic fire and other combustion products, smoke from agricultural burning, and soil dust created by overgrazing, deforestation, draining of inland water bodies, some farming practices, and generally, land management activities that destabilize the surface regolith to wind erosion. The amount of aerosol in the atmosphere has greatly increased in some parts of the world during the industrial period, and the nature of this particulate matter has substantially changed as a consequence of the evolving nature of emissions from industrial, commercial, agricultural, and residential activities, mainly combustion-related.

One of the greatest challenges in studying aerosol impacts on climate is the immense diversity, not only in particle size, composition, and origin, but also in spatial and temporal distribution. For most aerosols, whose primary source is emissions near the surface, concentrations are greatest in the atmospheric boundary layer, decreasing with altitude in the free troposphere. However, smoke from wildfires and volcanic effluent can be injected above the boundary layer; after injection, any type of aerosol can be lofted to higher elevations; this can extend their atmospheric lifetimes, increasing their impact spatially and climatically.

Aerosols are removed from the atmosphere primarily through cloud processing and wet

Figure 1.1. Major aerosol processes relevant to their impact on climate. Aerosols can be directly emitted as primary particles and can form secondarily by the oxidation of emitted gaseous precursors. Changes in relative humidity (RH) can cause particle growth or evaporation, and can alter particle properties. Physical processes within clouds can further alter particle properties, and conversely, aerosols can affect the properties of clouds, serving as condensation nuclei for new cloud droplet formation. Aqueous-phase chemical reactions in cloud drops or in clear air can also affect aerosol properties. Particles are ultimately removed from the atmosphere, scavenged by falling raindrops or settling by dry deposition. Modified from Ghan and Schwartz (2007).

Table 1.1. Estimated source strengths, lifetimes, mass loadings, and optical depths of major aerosol types. Statistics are based on results from 16 models examined by the Aerosol Comparisons between Observations and Models (AeroCom) project (Textor et al., 2006; Kinne et al., 2006). BC = black carbon; POM = particulate organic matter. See Chapter 3 for more details.

Aerosol Type	Total source[1] (Tg/yr[1])	Lifetime (day)	Mass loading[1] (Tg)	Optical depth @ 550 nm
	Median (Range)	Median (Range)	Median (Range)	Median (Range)
Sulfate[2]	190 (100-230)	4.1 (2.6-5.4)	2.0 (0.9-2.7)	0.034 (0.015-0.051)
BC	11 (8-20)	6.5 (5.3-15)	0.2 (0.05-0.5)	0.004 (0.002-0.009)
POM[2]	100 (50-140)	6.2 (4.3-11)	1.8 (0.5-2.6)	0.019 (0.006-0.030)
Dust	1600 (700-4000)	4.0 (1.3-7)	20 (5-30)	0.032 (0.012-0.054)
Sea salt	6000 (2000-120000)	0.4 (0.03-1.1)	6 (3-13)	0.030 (0.020-0.067)
Total				**0.13 (0.065-0.15)**

[1] Tg (teragram) = 10^{12} g, or million metric tons.

[2] The sulfate aerosol source is mainly SO_2 oxidation, plus a small fraction of direct emission. The organic matter source includes direct emission and hydrocarbon oxidation.

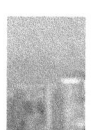

deposition in precipitation, a mechanism that establishes average tropospheric aerosol atmospheric lifetimes at a week or less (Table 1.1). The efficiency of removal therefore depends on the proximity of aerosols to clouds. For example, explosive volcanoes occasionally inject large amounts of aerosol precursors into the stratosphere, above most clouds; sulfuric acid aerosols formed by the 1991 Pinatubo eruption exerted a measurable effect on the atmospheric heat budget for several years thereafter (e.g., Minnis et al., 1993; McCormick et al., 1995; Robock, 2000, 2002). Aerosols are also removed by dry deposition processes: gravitational settling tends to eliminate larger particles, impaction typically favors intermediate-sized particles, and coagulation is one way smaller particles can aggregate with larger ones, leading to their eventual deposition by wet or dry processes. Particle injection height, subsequent air mass advection, and other factors also affect the rate at which dry deposition operates.

Despite relatively short average residence times, aerosols regularly travel long distances. For example, particles moving at mean velocity of 5 m s[-1] and remaining in the atmosphere for a week will travel 3000 km. Global aerosol observations from satellites provide ample evidence of this– Saharan dust reaches the Caribbean and Amazon basin, Asian desert dust and anthropogenic aerosol is found over the central Pacific and sometimes as far away as North America, and Siberian smoke can be deposited

in the Arctic. This transport, which varies both seasonally and inter-annually, demonstrates the global scope of aerosol influences.

As a result of the non-uniform distribution of aerosol sources and sinks, the short atmospheric lifetimes and intermittent removal processes compared to many atmospheric greenhouse trace gases, the spatial distribution of aerosol particles is quite non-uniform. The amount and nature of aerosols vary substantially with location and from year to year, and in many cases exhibit strong seasonal variations.

One consequence of this heterogeneity is that the impact of aerosols on climate must be understood and quantified on a *regional* rather than just a global-average basis. AOD trends observed in the satellite and surface-based data records suggest that since the mid-1990s, the amount of anthropogenic aerosol has decreased over North America and Europe, but has increased over parts of east and south Asia; on average, the atmospheric concentration of low-latitude smoke particles has increased (Mishchenko and Geogdzhayev, 2007). The observed AOD trends in the northern hemisphere are qualitatively consistent with changes in anthropogenic emissions (e.g. Streets et al., 2006a), and with observed trends in surface solar radiation flux ("solar brightening" or "dimming"), though other factors could be involved (e.g., Wild et al., 2005). Similarly, the increase in smoke parallels is associated with

The impact of aerosols on climate must be understood and quantified on a regional rather than just a global-average basis.

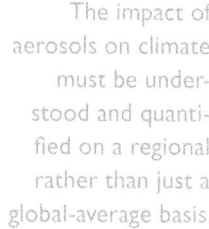

changing biomass burning patterns (e.g., Koren et al., 2007a).

1.2 The Climate Effects of Aerosols

Aerosols exert a variety of impacts on the environment. Aerosols (sometimes referred to as particulate matter or "PM," especially in air quality applications), when concentrated near the surface, have long been recognized as affecting pulmonary function and other aspects of human health. Sulfate and nitrate aerosols play a role in acidifying the surface downwind of gaseous sulfur and odd nitrogen sources. Particles deposited far downwind might fertilize iron-poor waters in remote oceans, and Saharan dust reaching the Amazon Basin is thought to contribute nutrients to the rainforest soil.

Aerosols also interact strongly with solar and terrestrial radiation in several ways. Figure 1.2 offers a schematic overview. First, they scatter and absorb sunlight (McCormick and Ludwig, 1967; Charlson and Pilat, 1969; Atwater, 1970; Mitchell, Jr., 1971; Coakley et al., 1983); these are described as "direct effects" on shortwave (solar) radiation. Second, aerosols act as sites at which water vapor can accumulate during cloud droplet formation, serving as cloud condensation nuclei or CCN. Any change in number concentration or hygroscopic properties of such particles has the potential to modify the physical and radiative properties of clouds,

altering cloud brightness (Twomey, 1977) and the likelihood and intensity with which a cloud will precipitate (e.g., Gunn and Phillips, 1957; Liou and Ou 1989; Albrecht, 1989). Collectively changes in cloud processes due to anthropogenic aerosols are referred to as *aerosol indirect effects*. Finally, absorption of solar radiation by particles is thought to contribute to a reduction in cloudiness, a phenomenon referred to as the *semi-direct effect*. This occurs because absorbing aerosol warms the atmosphere, which changes the atmospheric stability, and reduces surface flux.

The primary direct effect of aerosols is a brightening of the planet when viewed from space, as much of Earth's surface is dark ocean, and most aerosols scatter more than 90% of the visible light reaching them. The primary indirect effects of aerosols on clouds include an increase in cloud brightness, change in precipitation and possibly an increase in lifetime; thus the overall net impact of aerosols is an enhancement of Earth's reflectance (shortwave albedo). This reduces the sunlight reaching Earth's surface, producing a net climatic cooling, as well as a redistribution of the radiant and latent heat energy deposited in the atmosphere. These effects can alter atmospheric circulation and the water cycle, including precipitation patterns, on a variety of length and time scales (e.g., Ramanathan et al., 2001a; Zhang et al., 2006).

> The primary direct effect of aerosols is a brightening of the planet when viewed from space. The primary indirect effects of aerosols on clouds include an increase in cloud brightness and possibly an increase in lifetime. The overall net impact of aerosols is an enhancement of Earth's reflectance.

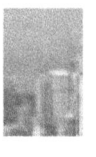

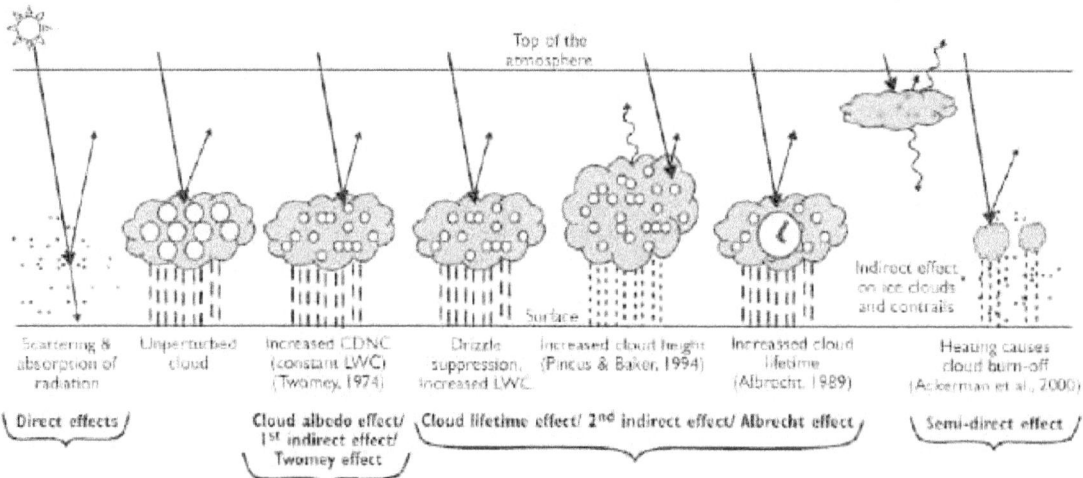

Figure 1.2. Aerosol radiative forcing. Airborne particles can affect the heat balance of the atmosphere, directly, by scattering and absorbing sunlight, and indirectly, by altering cloud brightness and possibly lifetime. Here small black dots represent aerosols, circles represent cloud droplets, straight lines represent short-wave radiation, and wavy lines, long-wave radiation. LWC is liquid water content, and CDNC is cloud droplet number concentration. Confidence in the magnitudes of these effects varies considerably (see Chapter 3). Although the overall effect of aerosols is a net cooling at the surface, the heterogeneity of particle spatial distribution, emission history, and properties, as well as differences in surface reflectance, mean that the magnitude and even the sign of aerosol effects vary immensely with location, season and sometimes inter-annually. The human-induced component of these effects is sometimes called "climate forcing." (From IPCC, 2007, modified from Haywood and Boucher, 2000).)

Several variables are used to quantify the impact aerosols have on Earth's energy balance; these are helpful in describing current understanding, and in assessing possible future steps.

For the purposes of this report, *aerosol radiative forcing* (RF) is defined as the net energy flux (downwelling minus upwelling) difference between an initial and a perturbed aerosol loading state, at a specified level in the atmosphere. (Other quantities, such as solar radiation, are assumed to be the same for both states.) This difference is defined such that a negative aerosol forcing implies that the change in aerosols relative to the initial state exerts a cooling influence, whereas a positive forcing would mean the change in aerosols exerts a warming influence.

There are a number of subtleties associated with this definition:

(1) The initial state against which aerosol forcing is assessed must be specified. For direct aerosol radiative forcing, it is sometimes taken as the complete absence of aerosols. IPCC AR4 (2007) uses as the initial state their estimate of aerosol loading in 1750. That year is taken as the approximate beginning of the era when humans exerted accelerated influence on the environment.

(2) A distinction must be made between aerosol RF and the *anthropogenic contribution* to aerosol RF. Much effort has been made to distinguishing these contributions by modeling and with the help of space-based, airborne, and surface-based remote sensing, as well as *in situ* measurements. These efforts are described in subsequent chapters.

(3) In general, aerosol RF and anthropogenic aerosol RF include energy associated with both the shortwave (solar) and the long-wave (primarily planetary thermal infrared) components of Earth's radiation budget. However, the solar component typically dominates, so in this document, these terms are used to refer to the solar component only, unless specified otherwise. The wavelength separation between the short- and long-wave components is usually set at around three or four micrometers.

(4) The IPCC AR4 (2007) defines radiative forcing as the net downward minus upward irradiance at the tropopause due to an external driver of climate change. This definition excludes stratospheric contributions to the overall forcing. Under typical conditions, most aerosols are located within the troposphere, so aerosol forcing at TOA and at the tropopause are expected to be very similar. Major volcanic eruptions or conflagrations can alter this picture regionally, and even globally.

(5) Aerosol radiative forcing can be evaluated at the surface, within the atmosphere, or at top-of-atmosphere (TOA). In this document, unless specified otherwise, aerosol radiative forcing is assessed at TOA.

(6) As discussed subsequently, aerosol radiative forcing can be greater at the surface than at TOA if the aerosols absorb solar radiation. TOA forcing affects the radiation budget of the planet. Differences between TOA forcing and surface forcing represent heating within the atmosphere that can affect vertical stability, circulation on many scales, cloud formation, and precipitation, all of which are climate effects of aerosols. In this document, unless specified otherwise, these additional climate effects are not included in aerosol radiative forcing.

(7) Aerosol direct radiative forcing can be evaluated under cloud-free conditions or under natural conditions, sometimes termed "all-sky" conditions, which include clouds. Cloud-free direct aerosol forcing is more easily and more accurately calculated; it is generally greater than all-sky forcing because clouds can mask the aerosol contribution to the scattered light. Indirect forcing, of course, must be evaluated for cloudy or all-sky conditions. In this document, unless specified otherwise, aerosol radiative forcing is assessed for all-sky conditions.

(8) Aerosol radiative forcing can be evaluated instantaneously, daily (24-hour) averaged, or assessed over some other time period. Many measurements, such as those from polar-orbiting satellites, provide instantaneous values, whereas models usually consider aerosol RF as a daily average quantity. In this document, unless specified otherwise, daily averaged aerosol radiative forcing is reported.

(9) Another subtlety is the distinction between a "forcing" and a "feedback." As different parts of the climate system interact, it is often unclear

Aerosol radiative forcing is defined as the net energy flux (downwelling minus upwelling) difference between an initial and a perturbed aerosol loading state.

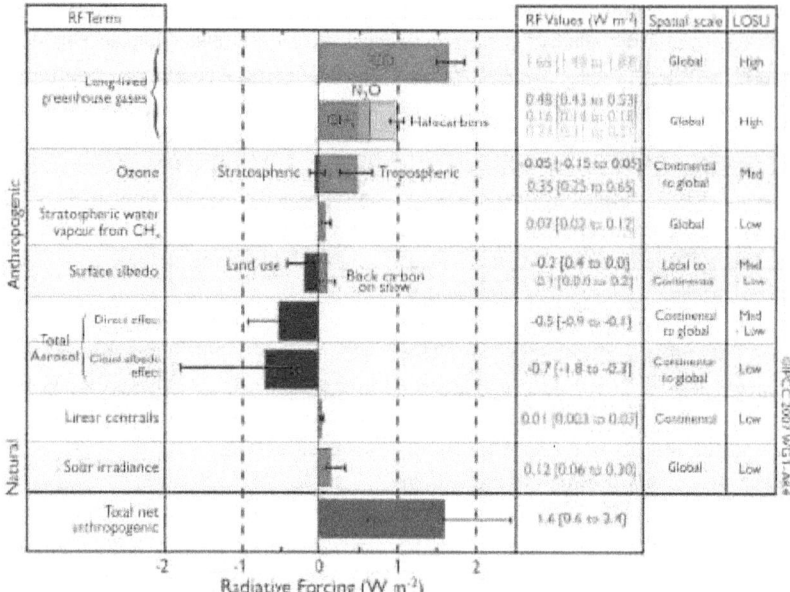

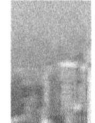

Figure I.3a. (Above) Global average radiative forcing (RF) estimates and uncertainty ranges in 2005, relative to the pre-industrial climate. Anthropogenic carbon dioxide (CO_2), methane (CH_4), nitrous oxide (N_2O), ozone, and aerosols as well as the natural solar irradiance variations are included. Typical geographical extent of the forcing (spatial scale) and the assessed level of scientific understanding (LOSU) are also given. Forcing is expressed in units of watts per square meter ($W m^{-2}$). The total anthropogenic radiative forcing and its associated uncertainty are also given. Figure from IPCC (2007).

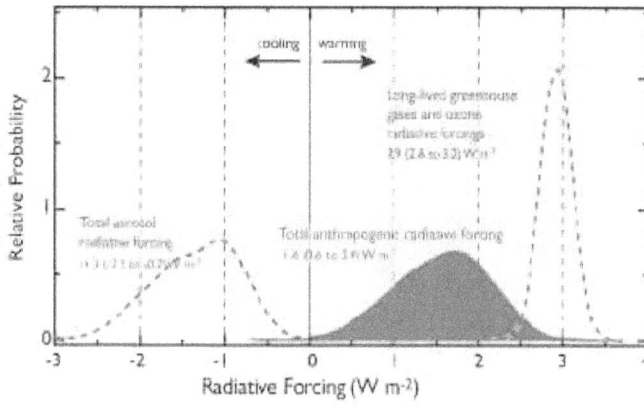

Figure I.3b. (Left) Probability distribution functions (PDFs) for anthropogenic aerosol and GHG RFs. Dashed red curve: RF of long-lived greenhouse gases plus ozone; dashed blue curve: RF of aerosols (direct and cloud albedo RF); red filled curve: combined anthropogenic RF. The RF range is at the 90% confidence interval. Figure adapted from IPCC (2007).

which elements are "causes" of climate change (forcings among them), which are responses to these causes, and which might be some of each. So, for example, the concept of aerosol effects on clouds is complicated by the impact clouds have on aerosols; the aggregate is often called aerosol-cloud interactions. This distinction sometimes matters, as it is more natural to attribute responsibility for causes than for responses. However, practical environmental considerations usually depend on the net result of all influences. In this report, "feedbacks" are taken as the consequences of changes in surface or atmospheric temperature, with the understanding that for some applications, the accounting may be done differently.

In summary, aerosol radiative forcing, the fundamental quantity about which this report is written, must be qualified by specifying the initial and perturbed aerosol states for which the radiative flux difference is calculated, the altitude at which the quantity is assessed, the wavelength regime considered, the temporal averaging, the cloud conditions, and whether total or only human-induced contributions are considered. The definition given here, qualified as needed, is used throughout the report.

Although the possibility that aerosols affect climate was recognized more than 40 years ago, the measurements needed to establish the magnitude of such effects, or even whether specific aerosol types warm or cool the surface, were lacking. Satellite instruments capable of at least crudely monitoring aerosol amount globally were first deployed in the late 1970s. But scientific focus on this subject grew substantially in the 1990s (e.g. Charlson et al., 1990; 1991; 1992; Penner et al., 1992), in part because it was recognized that reproducing the observed temperature trends over the industrial period with climate models requires including net global cooling by aerosols in the calculation (IPCC, 1995; 1996), along with the warming influence of enhanced atmospheric greenhouse gas (GHG) concentrations – mainly carbon dioxide, methane, nitrous oxide, chlorofluorocarbons, and ozone.

Improved satellite instruments, ground- and ship-based surface monitoring, more sophisticated chemical transport and climate models, and field campaigns that brought all these elements together with aircraft remote sensing and *in situ* sampling for focused, coordinated study, began to fill in some of the knowledge gaps. By the Fourth IPCC Assessment Report, the scientific community consensus held that in global average, the sum of direct and indirect top-of-atmosphere (TOA) forcing by anthropogenic aerosols is negative (cooling) of about -1.3 W m^{-2} (-2.2 to -0.5 W m^{-2}). This is significant compared to the positive forcing by anthropogenic GHGs (including ozone), about 2.9 ± 0.3 W m^{-2} (IPCC, 2007). However, the spatial distribution of the gases and aerosols are very different, and they do not simply exert compensating influences on climate.

The IPCC aerosol forcing assessments are based largely on model calculations, constrained as much as possible by observations. At present, aerosol influences are not yet quantified adequately, according to Figure 1.3a, as scientific understanding is designated as "Medium - Low" and "Low" for the direct and indirect climate forcing, respectively. The IPCC AR4 (2007) concluded that uncertainties associated with changes in Earth's radiation budget due to anthropogenic aerosols make the largest contribution to the overall uncertainty in radiative forcing of climate change among the factors assessed over the industrial period (Figure 3b).

Although AOD, aerosol properties, aerosol vertical distribution, and surface reflectivity all contribute to aerosol radiative forcing, AOD

usually varies on regional scales more than the other aerosol quantities involved. *Forcing efficiency (E_τ)*, defined as a ratio of direct aerosol radiative forcing to AOD at 550 nm, reports the sensitivity of aerosol radiative forcing to AOD, and is useful for isolating the influences of particle properties and other factors from that of AOD. E_τ is expected to exhibit a range of values globally, because it is governed mainly by aerosol size distribution and chemical composition (which determine aerosol single-scattering albedo and phase function), surface reflectivity, and solar irradiance, each of which exhibits pronounced spatial and temporal variations. To assess aerosol RF, E_τ is multiplied by the ambient AOD.

Figure 1.4 shows a range of E_τ, derived from AERONET surface sun photometer network measurements of aerosol loading and particle properties, representing different aerosol and surface types, and geographic locations. It demonstrates how aerosol direct solar radiative forcing (with initial state taken as the absence of aerosol) is determined by a combination of aerosol and surface properties. For example, E_τ due to southern African biomass burning smoke is greater at the surface and smaller at TOA than South American smoke because the southern African smoke absorbs sunlight more strongly, and the magnitude of E_τ for mineral dust for several locations varies depending on the underlying surface reflectance. Figure 1.4 illustrates one further point, that the radiative forcing by aerosols on surface energy balance can be much greater than that at TOA. This is especially true

Reproducing the observed temperature trends over the industrial period with climate models requires including net global cooling by aerosols in the calculation.

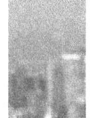

The radiative forcing by aerosols on surface energy balance can be much greater than that at the top of the atmosphere.

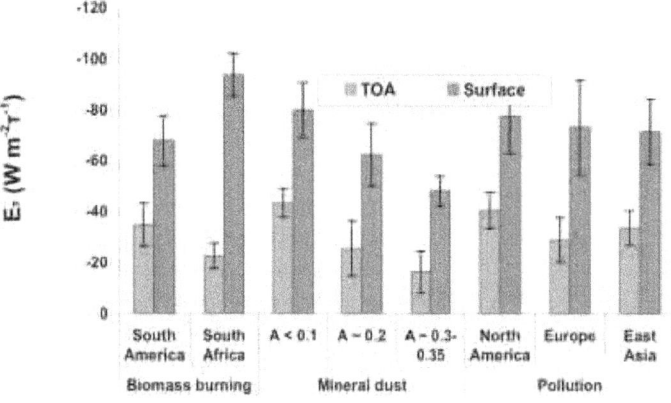

Figure 1.4. The clear-sky forcing efficiency E_τ, defined as the diurnally averaged aerosol direct radiative effect (W m^{-2}) per unit AOD at 550 nm, calculated at both TOA and the surface, for typical aerosol types over different geographical regions. The vertical black lines represent ± one standard deviation of E_τ for individual aerosol regimes and A is surface broadband albedo. (adapted from Zhou et al., 2005).

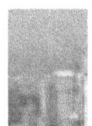

In regions having high concentrations of anthropogenic aerosol, aerosol forcing is much stronger than the global average, and can exceed the magnitude of greenhouse gas warming.

when the particles have SSA substantially less than 1, which can create differences between surface and TOA forcing as large as a factor of five (e.g., Zhou et al., 2005).

Table 1.2 presents estimates of cloud-free, instantaneous, aerosol direct RF dependence on AOD, and on aerosol and surface properties, calculated for three sites maintained by the US Department of Energy's Atmospheric Radiation Measurement (ARM) program, where surface and atmospheric conditions span a significant range of natural environments (McComiskey et al., 2008a). Here aerosol RF is evaluated relative to an initial state that is the complete absence of aerosols. Note that aerosol direct RF dependence on individual parameters varies considerably, depending on the values of the other parameters, and in particular, that aerosol RF dependence on AOD actually changes sign, from net cooling to net warming, when aerosols reside over an exceedingly bright surface. Sensitivity values are given for snapshots at fixed solar zenith angles, relevant to measurements made, for example, by polar-orbiting satellites.

The lower portion of Table 1.2 presents upper bounds on instantaneous measurement uncertainty, assessed individually for each of AOD, SSA, g, and A, to produce a 1 W m^{-2} top-of-atmosphere, cloud-free aerosol RF accuracy. The values are derived from the upper portion of the table, and reflect the diversity of conditions captured by the three ARM sties. Aerosol RF sensitivity of 1 W m^{-2} is used as an example; uncertainty upper bounds are obtained from the partial derivative for each parameter by neglecting the uncertainties for all other parameters. These estimates produce an instantaneous AOD measurement uncertainty upper bound between about 0.01 and 0.02, and SSA constrained to about 0.02 over surfaces as bright or brighter than the ARM Southern Great Plains site, typical of mid-latitude, vegetated land. Other researchers, using independent data sets, have derived ranges of E_τ and aerosol RF sensitivity similar to those presented here, for a variety of conditions (e.g., Christopher and Jones, 2008; Yu et al., 2006; Zhou et al., 2005).

These uncertainty bounds provide a baseline against which current and expected near-future instantaneous measurement capabilities are assessed in Chapter 2. Model sensitivity is usually evaluated for larger-scale (even global)

Radiative heating of the atmosphere by absorbing particles can change the atmospheric temperature structure, affecting vertical mixing, cloud formation and evolution, and possibly large-scale dynamical systems.

and longer-term averages. When instantaneous measured values from a randomly sampled population are averaged, the uncertainty component associated with random error diminishes as something like the inverse square root of the number of samples. As a result, the accuracy limits used for assessing more broadly averaged model results corresponding to those used for assessing instantaneous measurements, would have to be tighter, as discussed in Chapter 4.

In summary, much of the challenge in quantifying aerosol influences arises from large spatial and temporal heterogeneity, caused by the wide variety of aerosol sources, sizes and compositions, the spatial non-uniformity and intermittency of these sources, the short atmospheric lifetime of most aerosols, and the spatially and temporally non-uniform chemical and microphysical processing that occurs in the atmosphere. In regions having high concentrations of anthropogenic aerosol, for example, aerosol forcing is much stronger than the global average, and can exceed the magnitude of GHG warming, locally reversing the sign of the net forcing. It is also important to recognize that the global-scale aerosol TOA forcing alone is not an adequate metric for climate change (NRC, 2005). Due to aerosol absorption, mainly by soot, smoke, and some desert dust particles, the aerosol direct radiative forcing at the surface can be much greater than the TOA forcing, and in addition, the radiative heating of the atmosphere by absorbing particles can change the atmospheric temperature structure, affecting vertical mixing, cloud formation and evolution, and possibly large-scale dynamical systems such as the monsoons (Kim et al., 2006; Lau et al., 2008). By realizing aerosol's climate significance and the challenge of charactering highly variable aerosol amount and properties, the US Climate Change Research Initiative (CCRI) identified research on atmospheric concentrations and effects of aerosols specifically as a top priority (NRC, 2001).

1.3. Reducing Uncertainties in Aerosol-Climate Forcing Estimates

Regional as well as global aerosol radiative effects on climate are estimated primarily through the use of climate models (e.g., Penner et al., 1994; Schulz et al., 2006). These numerical models are evaluated based on their ability to simulate the aerosol- and cloud-related processes that affect climate for current and past

Table 1.2. Top-of-atmosphere, cloud-free, instantaneous direct aerosol radiative forcing dependence on aerosol and surface properties. Here TWP, SGP, and NSA are the Tropical West Pacific island, Southern Great Plains, and North Slope Alaska observation stations maintained by the DOE ARM program, respectively. Instantaneous values are given at specific solar zenith angle. Upper and middle parts are from McComiskey et al. (2008a). Representative, parameter-specific measurement uncertainty upper bounds for producing 1 W m^{-2} instantaneous TOA forcing accuracy are given in the lower part, based on sensitivities at three sites from the middle part of the table.

Parameters	TWP	SGP	NSA
Aerosol properties (AOD, SSA, g), solar zenith angle (SZA), surface albedo (A), and aerosol direct RF at TOA (F):			
AOD	0.05	0.1	0.05
SSA	0.97	0.95	0.95
g	0.8	0.6	0.7
A	0.05	0.1	0.9
SZA	30	45	70
F (W m^{-2})	-2.2	-6.3	2.6
Sensitivity of cloud-free, instantaneous, TOA direct aerosol radiative forcing to aerosol and surface properties , W m^{-2} per unit change in property:			
$\partial F/\partial$(AOD)	-45	-64	51
$\partial F/\partial$(SSA)	-11	-50	-60
$\partial F/\partial g$	13	23	2
$\partial F/\partial A$	8	24	6
Representative measurement uncertainty upper bounds for producing 1 W m^{-2} accuracy of aerosol RF:			
AOD	0.022	0.016	0.020
SSA	0.091	0.020	0.017
g	0.077	0.043	
A	0.125	0.042	0.167

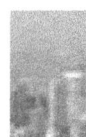

conditions. The derived accuracy serves as a measure of the accuracy with which the models might be expected to predict the dependence of future climate conditions on prospective human activities. To generate such predictions, the models must simulate the physical, chemical, and dynamical mechanisms that govern aerosol formation and evolution in the atmosphere (Figure 1.1), as well as the radiative processes that govern their direct and indirect climate impact (Figure 1.2), on all the relevant space and time scales.

Some models simulate aerosol emissions, transports, chemical processing, and sinks, using atmospheric and possibly also ocean dynamics generated off-line by separate numerical systems. These are often called Chemistry and Transport Models (CTMs). In contrast, General Circulation Models or Global Climate Models (GCMs) can couple aerosol behavior and dynamics as part of the same calculation, and are capable of representing interactions between aerosols and dynamical aspects of the climate system, although currently many of them still use prescribed aerosols to study climate sensitivity.

The IPCC AR4 total anthropogenic radiative forcing estimate, shown in Figure 1.3, is 1.6 W m^{-2} from preindustrial times to the present, with a likely range of 0.6 to 2.4 W m^{-2}. This estimate includes long-lived GHGs, ozone, and aerosols. The increase in global mean surface temperature of 0.7°C, from the transient climate simulations in response to this forcing, yields a transient climate sensitivity (defined as the surface temperature change per unit RF) over the industrial period of 0.3 to 1.1°C/(W m^{-2}).

Under most emission scenarios, CO_2 is expected to double by the latter part of the 21st century. A climate sensitivity range of 0.3 to 1.1°C/(W m^{-2}) translates into a future surface temperature increase attributable to CO_2 forcing at the time of doubled CO_2 of 1.2 to 4.7°C. Such a range is too wide to meaningfully predict the climate response to increased greenhouse gases (e.g., Caldeira et al., 2003). As Figure 1.3 shows, the largest contribution to overall uncertainty in estimating the climate response is from aerosol RF.

The key to reducing uncertainty in the role of aerosols in climate is to much better represent the processes that contribute to the aerosol climate effects in models. This report highlights three specific areas for continued, focused effort: (1) improving measurement quality and coverage, (2) achieving more effective use of measurements to constrain model simulations and to test model parameterizations, and (3) producing more accurate representation of aerosols and clouds in models. This section provides a brief introduction to the current state of aerosol measurements and model representations of aerosol processes, as they relate to assessing aerosol impacts on climate. More complete discussion of these topics and assessment of possible next steps are given in Chapters 2, 3, and 4.

Improving measurement quality and coverage. Aerosol mass concentration, size and composition distributions, and absorption properties, as functions of location and time, are the main aerosol-specific elements of CTMs. They depend on primary particle and precursor gas emissions, on gas-to-particle conversion processes, on transport, humidification and cloud processing, and removal mechanisms. Satellite instruments, surface-based networks (*in situ* and remote sensing), and research aircraft all contribute quantitative measurements of aerosol properties and/or distributions that can be used to help constrain models, as well as to test and refine the model representations of processes that govern aerosol life cycles. As described in Chapter 2, the current situation reflects the significant progress that has been made over the past decade in satellite, airborne, ground-based and laboratory instrumentation, actual measurements available from each of these sources, remote sensing retrieval methods, and data validation techniques.

However, each type of measurement is limited in terms of the accuracy, and spatial and temporal sampling of measured quantities. At present, satellite passive imagers monitor AOD globally up to once per day, with accuracies under cloud-free, good but not necessarily ideal viewing conditions of about 0.05 or (0.1 to 0.2) x AOD, whichever is larger, for vegetated land, somewhat better over dark water, and less well over bright desert (e.g., Kahn et al., 2005a; Remer et al., 2005). Reliable AOD retrieval over snow and ice from passive remote sensing imagers has not yet been achieved. From space, aerosol vertical distribution is provided mainly by lidars that offer sensitivity to multiple layers, even in the presence of thin cloud, but they require several weeks to observe just a fraction of a percent of the planet.

From the expansive vantage point of space, there is enough information to identify column-average ratios of coarse to fine AOD, or even aerosol air mass types in some circumstances, but not sufficient to deduce chemical composition and vertical distribution of type, nor to constrain light absorption approaching the ~0.02 SSA sensitivity suggested in Section 1.2.

As a result, it is difficult to separate anthropogenic from natural aerosols using currently available satellite data alone, though attempts at this have been made based on retrieved particle size and shape information (see Chapter 2). At present, better quantification of anthropogenic aerosol depends upon integrating satellite measurements with other observations and models. Aircraft and ground-based *in situ* sampling can help fill in missing physical and chemical detail, although coverage is very limited in both space and time. Models can contribute by connecting observed aerosol distributions with likely sources and associated aerosol types. Surface remote-sensing monitoring networks offer temporal resolution of minutes to hours, and greater column AOD accuracy than satellite observations, but height-resolved particle property information has been demonstrated by only a few cutting-edge technologies such as high-spectral-resolution lidar (HSRL), and again, spatial coverage is extremely limited.

Even for satellite observations, sampling is an issue. From the passive imagers that provide the greatest coverage, AOD retrievals can only be done under cloud-free conditions, leading to a "clear-sky bias," and there are questions about retrieval accuracy in the vicinity of clouds. And retrievals of aerosol type from these instruments as well as from surface-based passive remote sensing require at least a certain minimum column AOD to be effective; the thresholds depend in part on aerosol type itself and on surface reflectivity, leading to an "AOD bias" in these data sets.

Other measurement-related issues include obtaining sufficiently extensive aerosol vertical distributions outside the narrow sampling beam of space-based, airborne, or ground-based lidars, retrieving layer-resolved aerosol properties, which is especially important in the many regions where multiple layers of different types are common, obtaining representative *in situ* samples of large particles, since they tend to be under-sampled when collected by most aircraft inlets, and acquiring better surface measurement coverage over oceans.

Achieving more effective use of measurements to constrain models. Due to the limitations associated with each type of observational data record, reducing aerosol-forcing uncertainties requires coordinated efforts at integrating data from multiple platforms and techniques (Seinfeld et al., 1996; Kaufman et al., 2002a; Diner et al., 2004; Anderson et al., 2005a). Initial steps have been taken to acquire complementary observations from multiple platforms, especially through intensive field campaigns, and to merge data sets, exploiting the strengths of each to provide better constraints on models (e.g., Bates et al., 2006; Yu et al., 2006; Kinne et al., 2006; see Chapter 2, Section 2.2.6). Advanced instrument concepts, coordinated measurement strategies, and retriev-

al techniques, if implemented, promise to further improve the contributions observations make to reducing aerosol forcing uncertainties.

Producing more accurate representation of aerosols in models. As discussed in Chapter 3, models, in turn, have developed increasingly sophisticated representations of aerosol types and processes, have improved the spatial resolution at which simulations are performed, and through controlled experiments and inter-comparisons of results from many models, have characterized model diversity and areas of greatest uncertainty (e.g., Textor et al., 2006; Kinne et al., 2006).

A brief chronology of aerosol modeling used for the IPCC reports illustrates these developments. In the IPCC First Assessment Report (1990), the few transient climate change simulations that were discussed used only increases in greenhouse gases. By IPCC Second Assessment Report (1995), although most GCMs still considered only greenhouse gases, several simulations included the direct effect of sulfate aerosols. The primary purpose was to establish whether the pattern of warming was altered by including aerosol-induced cooling in regions of high emissions such as the Eastern U.S. and eastern Asia. In these models, the sulfate aerosol distribution was derived from a sulfur cycle model constrained by estimated past aerosol emissions and an assumed future sulfur emission scenario. The aerosol forcing contribution was mimicked by increasing the surface albedo, which improved model agreement with the observed global mean temperature record for the final few decades of the twentieth century, but not for the correct reasons (see Chapter 3).

The IPCC Third Assessment Report (TAR, 2001) report cited numerous groups that included aerosols in both 20th and 21st century simulations. The direct effect of sulfate aerosols was required to reproduce the observed global temperature change, given the models' climate sensitivity and ocean heat uptake. Although most models still represented aerosol forcing by increasing the surface albedo, several groups explicitly represented sulfate aerosols in their atmospheric scattering calculations, with geographical distributions determined by off-line CTM calculations. The first model calculations that included any indirect effects of aerosols on clouds were also presented.

Due to the limitations associated with each type of observational data record, reducing aerosol-forcing uncertainties requires coordinated efforts at integrating data from multiple platforms and techniques.

The most recent IPCC assessment report (AR4; 2007) summarized the climate change experiments from more than 20 modeling groups that this time incorporated representations of multiple aerosol species, including black and organic carbon, mineral dust, sea salt and in some cases nitrates (see Chapter 3). In addition, many attempts were made to simulate indirect effects, in part because the better understood direct effect appeared to be insufficient to properly simulate observed temperature changes, given model sensitivity. As in previous assessments, the AR4 aerosol distributions responsible for both the direct and indirect effect were produced off-line, as opposed to being run in a coupled mode that would allow simulated climate changes to feed back on the aerosol distributions.

The fact that models now use multiple aerosol types and often calculate both direct and indirect aerosol effects does not imply that the requisite aerosol amounts and optical characteristics, or the mechanisms of aerosol-cloud interactions, are well represented. For example, models tend to have lower AOD relative to measurements, and are poorly constrained with regard to speciation (see Table 3.2 and Figure 3.1 in Chapter 3). To bridge the gap between measurements and models in this area, robust relationships need to be established for different aerosol types, connecting the AOD and types retrieved from spacecraft, aircraft, and surface remote sensing observations, with the aerosol mass concentrations that are the fundamental aerosol quantities tracked in CTMs and GCMs.

As detailed below, continued progress with measurement, modeling, and at the interface between the two, promises to improve estimates of aerosol contributions to climate change, and to reduce the uncertainties in these quantities reflected in Figure 1.3.

1.4 Contents of This Report

This report assesses current understanding of aerosol radiative effects on climate, focusing on developments of aerosol measurement and modeling subsequent to IPCC TAR (2001). It reviews the present state of understanding of aerosol influences on Earth's climate system, and in particular, the consequences for climate change of their direct and indirect effects. This report does not deal with several natural forcings that involve aerosols. Stratospheric aerosols produced by large volcanic eruptions exert large, short-term effects which are particularly important for characterizing climate system response to forcing, and the effects of recent eruptions (e.g. Pinatubo) are well documented (e.g., Minnis et al., 1993; McCormick et al., 1995; Robock et al., 2002). However these effects are intermittent and have only short-term environmental impacts (ca. 1 year). Galactic cosmic rays, modulated by the 11-year solar cycle, have been reported to correlate with the total cloud cover (e.g., Svensmark and Friis-Christensen, 1997), possibly by aiding the nucleation of new particles that grow into cloud condensation nuclei (e.g., Turco et al., 1998). However, the present mainstream consensus is that these phenomena exert little to no effect on cloud cover or other cloud properties (e.g., Lockwood and Fröhlich, 2008; Kristjánsson et al., 2008).

The Executive Summary reviews the key concepts involved in the study of aerosol effects on climate, and provides a chapter-by-chapter summary of conclusions from this assessment. Chapter 1 provides basic definitions, radiative forcing accuracy requirements, and background material on critical issues needed to motivate the more detailed discussion and assessment given in subsequent chapters.

Chapter 2 assesses the aerosol contributions to radiative forcing based on remote sensing and *in situ* measurements of aerosol amounts and properties. Current measurement capabilities and limitations are discussed, as well as synergy with models, in the context of the needed aerosol radiative forcing accuracy.

Model simulation of aerosols and their direct and indirect effects are examined in Chapter 3. Representations of aerosols used for IPCC AR4 (2007) climate simulations are discussed, providing an overview of near-term modeling option strengths and limitations for assessing aerosol forcing of climate.

Finally, Chapter 4 provides an assessment of how current capabilities, and those within reach for the near future, can be brought together to reduce the aerosol forcing uncertainties reported in IPCC AR4 (2007).

Continued progress with measurement, modeling, and at the interface between the two, promises to improve estimates of aerosol contribution to climate change.

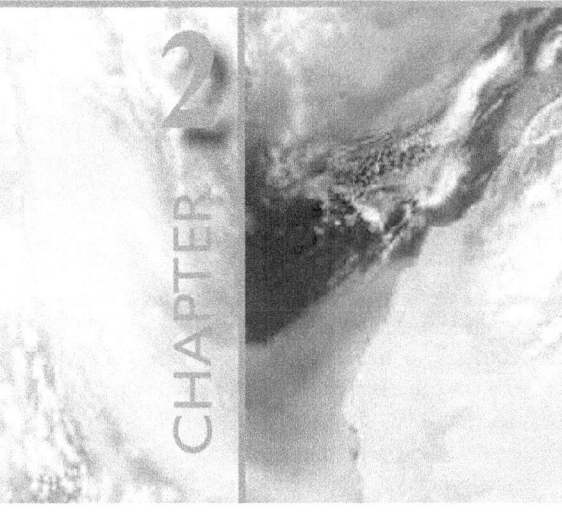

Remote Sensing and *In Situ* Measurements of Aerosol Properties, Burdens, and Radiative Forcing

Lead Authors: Hongbin Yu, NASA GSFC/UMBC; Patricia K. Quinn, NOAA PMEL; Graham Feingold, NOAA ESRL; Lorraine A. Remer, NASA GSFC; Ralph A. Kahn, NASA GSFC
Contributing Authors: Mian Chin, NASA GSFC; Stephen E. Schwartz, DOE BNL

2.1. Introduction

As discussed in Chapter 1, much of the challenge in quantifying aerosol direct radiative forcing (DRF) and aerosol-cloud interactions arises from large spatial and temporal heterogeneity of aerosol concentrations, compositions, and sizes, which requires an integrated approach that effectively combines measurements and model simulations. Measurements, both *in situ* and remote sensing, play essential roles in this approach by providing data with sufficient accuracy for validating and effectively constraining model simulations. For example, to achieve an accuracy of 1 W m^{-2} for the instantaneous, top-of-atmosphere (TOA) aerosol DRF under cloud free conditions, the accuracy for measuring aerosol optical depth (AOD) should be within 0.01 and 0.02 for mid-visible wavelength, and that for single-scattering albedo (SSA) should be constrained to about 0.02 over land (Chapter 1, Table 1.2). The measurement requirements would be much tighter in order to achieve the same forcing accuracy at the surface. Quantifying anthropogenic component of DRF and aerosol indirect radiative forcing would impose additional accuracy requirements on measurements of aerosol chemical composition and microphysical properties (e.g., size distribution) that are needed to attribute material to sources or source type.

Over the past decade and since the Intergovermental Panel on Climate Change (IPCC) Third Assessment Report (TAR) (IPCC 2001) in particular, a great deal of effort has gone into improving measurement data sets (as summarized in Yu et al., 2006; Bates et al., 2006; Kahn et al., 2004). Principal efforts have been:

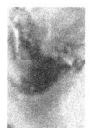

- Development and implementation of new and enhanced satellite-borne sensors examining aerosol effects on atmospheric radiation;
- Execution of focused field experiments examining aerosol processes and properties in various aerosol regimes around the globe;
- Establishment and enhancement of ground-based networks measuring aerosol properties and radiative forcing; and
- Development and deployment of new and enhanced instrumentation, importantly aerosol mass spectrometers examining size dependent composition and several methods for measuring aerosol SSA.

These efforts have made it feasible to shift the estimates of aerosol radiative forcing increasingly from largely model-based as in IPCC TAR to measurement-based as in the IPCC Fourth Assessment Report (AR4) (IPCC 2007). Satellite measurements that are evaluated, supplemented, and constrained by ground-based remote sensing measurements and *in situ* measurements from focused field campaigns, provide the basis for the regional- to global-scale assessments. Chemistry and transport models (CTMs) are used to interpolate and supplement the data in regions and under conditions where observational data are not available or to assimilate high-quality data from various observations to constrain and thereby improve model simulations of aerosol impacts. These developments have played an important role in advancing the scientific understanding of aerosol direct and indirect radiative forcing as documented in the IPCC AR4 (IPCC, 2007).

The goals of this chapter are to:
- provide an overview of current aerosol measurement capabilities and limitations;
- describe the concept of synergies between different types of measurements and models;
- assess estimates of aerosol direct and indirect radiative forcing from different observational approaches; and
- discuss outstanding issues to which measurements can contribute.

The synthesis and assessment in this chapter lays groundwork needed to develop a future research strategy for understanding and quantifying aerosol-climate interactions.

2.2. Overview of Aerosol Measurement Capabilities

2.2.1. SATELLITE REMOTE SENSING

A measurement-based characterization of aerosols on a global scale can be realized only through satellite remote sensing, which is the only means of characterizing the large spatial and temporal heterogeneities of aerosol distributions. Monitoring aerosols from space has been performed for over two decades and is planned for the coming decade with enhanced capabilities (King et al., 1999; Foster et al., 2007; Lee et al., 2006; Mishchenko et al., 2007b). Table 2.1 summarizes major satellite measurements currently available for the tropospheric aerosol characterization and radiative forcing research.

Early aerosol monitoring from space relied on sensors that were designed for other purposes. The Advanced Very High Resolution Radiometer (AVHRR), intended as a cloud and surface monitoring instrument, provides radiance observations in the visible and near infrared wavelengths that are sensitive to aerosol properties over the ocean (Husar et al., 1997; Mishchenko et al., 1999). Originally intended for ozone monitoring, the ultraviolet (UV) channels used for the Total Ozone Mapping Spectrometer (TOMS) are sensitive to aerosol UV absorption

with little surface interferences, even over land (Torres et al., 1998). This UV-technique makes TOMS suitable for monitoring biomass burning smoke and dust, though with limited sensitivity near the surface (Herman et al., 1997) and for retrieving aerosol single-scattering albedo from space (Torres et al., 2005). (A new sensor, the Ozone Monitoring Instrument (OMI) aboard Aura, has improved on such UV-technique advantages, providing higher spatial resolution and more spectral channels, see Veihelmann et al., 2007). Such historical sensors have provided multi-decadal climatology of aerosol optical depth that has significantly advanced the understanding of aerosol distributions and long-term variability (e.g., Geogdzhayev et al., 2002; Torres et al., 2002; Massie et al., 2004; Mishchenko et al., 2007a; Mishchenko and Geogdzhayev, 2007; Zhao et al., 2008a).

Over the past decade, satellite aerosol retrievals have become increasingly sophisticated. Now, satellites measure the angular dependence of radiance and polarization at multiple wavelengths from UV through the infrared (IR) at fine spatial resolution. From these observations, retrieved aerosol products include not only optical depth at one wavelength, but also spectral optical depth and some information about particle size over both ocean and land, as well as more direct measurements of polarization and phase function. In addition, cloud screening is much more robust than before and onboard calibration is now widely available. Examples of such new and enhanced sensors include the MODerate resolution Imaging Spectroradiometer (MODIS, see Box 2.1), the Multi-angle Imaging SpectroRadiometer (MISR, see Box 2.2), Polarization and Directionality of the Earth's Reflectance (POLDER, see Box 2.3), and OMI, among others. The accuracy for AOD measurement from these sensors is about 0.05 or 20% of AOD (Remer et al., 2005; Kahn et al., 2005a) and somewhat better over dark water, but that for aerosol microphysical properties, which is useful for distinguishing aerosol air mass types, is generally low. The Clouds and the Earth's Radiant Energy System (CERES, see Box 2.4) measures broadband solar and terrestrial radiances. The CERES radiation measurements in combination with satellite retrievals of aerosol optical depth can be used to determine aerosol direct radiative forcing.

Satellite remote sensing is the only means of characterizing the large spatial and temporal heterogeneities of aerosol distributions.

Table 2.1. Summary of major satellite measurements currently available for the tropospheric aerosol characterization and radiative forcing research.

Category	Properties	Sensor/platform	Parameters	Spatial coverage	Temporal coverage
Column-integrated	Loading	AVHRR/NOAA-series	optical depth	~daily coverage of global ocean	1981-present
		TOMS/Nimbus, ADEOSI, EP		~daily coverage of global land and ocean	1979-2001
		POLDER-1, -2, PARASOL			1997-present
		MODIS/Terra, Aqua			2000-present (Terra) 2002-present (Aqua)
		MISR/Terra		~weekly coverage of global land and ocean, including bright desert and nadir sun-glint	2000-present
		OMI/Aura		~daily coverage of global land and ocean	2005-present
	Size, shape	AVHRR/NOAA-series	Ångström exponent	global ocean	1981-present
		POLDER-1, -2, PARASOL	fine-mode fraction, Ångström exponent, non-spherical fraction	global land+ocean	1997-present
		MODIS/Terra, Aqua	fine-mode fraction	global land+ocean (better quality over ocean)	2000-present (Terra) 2002-present (Aqua)
			Ångström exponent		
			effective radius	global ocean	
			asymmetry factor		
		MISR/Terra	Ångström exponent, small, medium, large fractions, non-spherical fraction	global land+ocean	2000-present
	Absorption	TOMS/Nimbus, ADEOSI, EP	absorbing aerosol index, single-scattering albedo, absorbing optical depth	global land+ocean	1979-2001
		OMI/Aura			2005-present
		MISR/Terra	single-scattering albedo (2-4 bins)		2000-present
Vertical-resolved	Loading, size, and shape	GLAS/ICESat	extinction/backscatter	global land+ocean, 16-day repeating cycle, single-nadir measurement	2003-present (~3months/year)
		CALIOP/CALIPSO	extinction/backscatter, color ratio, depolarization ratio		2006-present

23

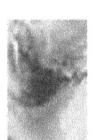

Box 2.1: MODerate resolution Imaging Spectroradiometer

MODIS performs near global daily observations of atmospheric aerosols. Seven of 36 channels (between 0.47 and 2.13 μm) are used to retrieve aerosol properties over cloud and surface-screened areas (Martins et al., 2002; Li et al., 2004). Over vegetated land, MODIS retrieves aerosol optical depth at three visible channels with high accuracy of ±0.05±0.2τ (Kaufman et al., 1997; Chu et al., 2002; Remer et al., 2005; Levy et al., 2007b). Most recently a deep-blue algorithm (Hsu et al., 2004) has been implemented to retrieve aerosols over bright deserts on an operational basis, with an estimated accuracy of 20-30%. Because of the greater simplicity of the ocean surface, MODIS has the unique capability of retrieving not only aerosol optical depth with greater accuracy, i.e., ±0.03±0.05τ (Tanré et al., 1997; Remer et al., 2002; 2005; 2008), but also quantitative aerosol size parameters (e.g., effective radius, fine-mode fraction of AOD) (Kaufman et al., 2002a; Remer et al., 2005; Kleidman et al., 2005). The fine-mode fraction has been used as a tool for separating anthropogenic aerosol from natural ones and estimating the anthropogenic aerosol direct climate forcing (Kaufman et al., 2005a). Figure 2.1 shows composites of MODIS AOD and fine-mode fraction that illustrate seasonal and geographical variations of aerosol types. Clearly seen from the figure is heavy pollution over East Asia in both months, biomass burning smoke over South Africa, South America, and Southeast Asia in August, heavy dust storms over North Atlantic in both months and over Arabian Sea in August, and a mixture of dust and pollution plume swept across North Pacific in April.

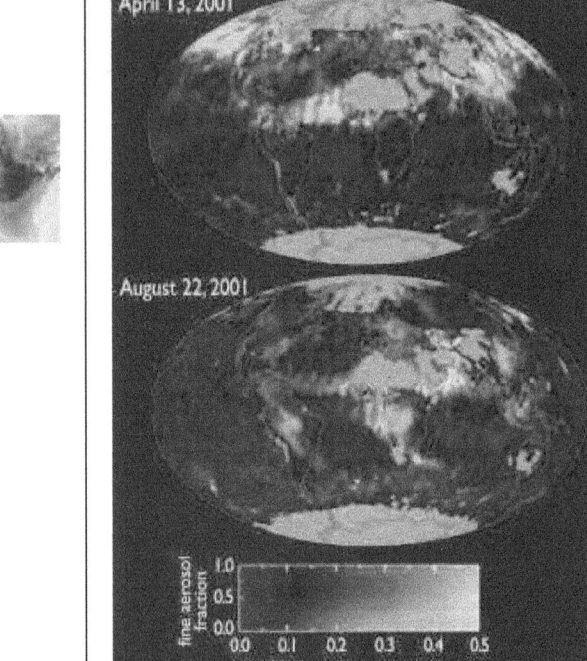

Figure 2.1. A composite of MODIS/Terra observed aerosol optical depth (at 550 nm, green light near the peak of human vision) and fine-mode fraction that shows spatial and seasonal variations of aerosol types. Industrial pollution and biomass burning aerosols are predominately small particles (shown as red), whereas mineral dust and sea salt consist primarily of large particles (shown as green). Bright red and bright green indicate heavy pollution and dust plumes, respectively (adapted from Chin et al., 2007; original figure from Yoram Kaufman and Reto Stöckli).

Complementary to these passive sensors, active remote sensing from space is also now possible and ongoing (see Box 2.5). Both the Geoscience Laser Altimeter System (GLAS) and the Cloud and Aerosol Lidar with Orthogonal Polarization (CALIOP) are collecting essential information about aerosol vertical distributions. Furthermore, the constellation of six afternoon-overpass spacecrafts (as illustrated in Figure 2.5), the so-called A-Train (Stephens et al., 2002) makes it possible for the first time to conduct near simultaneous (within 15-minutes) measurements of aerosols, clouds, and radiative

fluxes in multiple dimensions with sensors in complementary capabilities.

The improved accuracy of aerosol products (mainly AOD) from these new-generation sensors, together with improvements in characterizing the earth's surface and clouds, can help reduce the uncertainties associated with estimating the aerosol direct radiative forcing (Yu et al., 2006; and references therein). The retrieved aerosol microphysical properties, such as size, absorption, and non-spherical fraction can help distinguish anthropogenic aerosols from natural aerosols and hence help assess the anthropogenic component of aerosol direct radiative forcing (Kaufman et al., 2005a; Bellouin et al., 2005, 2008; Christopher et al., 2006; Yu et al., 2006, 2008). However, to infer aerosol number concentrations and examine indirect aerosol radiative effects from space, significant efforts are needed to measure aerosol size distribution with much improved accuracy, characterize aerosol type, account for impacts of water uptake on aerosol optical depth, and determine the fraction of aerosols that is at the level of the clouds (Kapustin et al., 2006; Rosenfeld, 2006). In addition, satellite remote sensing is not sensitive to particles much smaller than 0.1 micrometer in diameter, which comprise of a significant fraction of those that serve as cloud condensation nuclei.

Finally, algorithms are being developed to retrieve aerosol absorption or SSA from satellite observations (e.g., Kaufman et al., 2002b; Torres et al., 2005). The NASA Glory mission, scheduled to launch in 2009 and to be added to the A-Train, will deploy a multi-angle, multi-spectral polarimeter to determine the global distribution of aerosol and clouds. It will also be able to infer microphysical property information, from which aerosol type (e.g., marine, dust, pollution, etc.) can be inferred for improving quantification of the aerosol direct and indirect forcing on climate (Mishchenko et al., 2007b).

In summary, major advances have been made in both passive and active aerosol remote sensing from space in the past decade, providing better coverage, spatial resolution, retrieved AOD accuracy, and particle property information. However, AOD accuracy is still much poorer than that from surface-based sun photometers (0.01 to 0.02), even over vegetated land and dark water where retrievals are most reliable. Although there is some hope of approaching this level of uncertainty with a new generation of satellite instruments, the satellite retrievals entail additional sensitivities to aerosol and surface scattering properties. It seems unlikely that satellite remote sensing could exceed the sun photometer accuracy without introducing some

Major advances have been made in both passive and active aerosol remote sensing from space in the past decade, providing better coverage, spatial resolution, retrieved AOD accuracy, and particle property information.

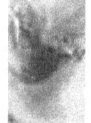

Box 2.2: Multi-angle Imaging SpectroRadiometer

MISR, aboard the sun-synchronous, polar orbiting satellite Terra, measures upwelling solar radiance in four visible-near-IR spectral bands and at nine view angles spread out in the forward and aft directions along the flight path (Diner et al., 2002). It acquires global coverage about once per week. A wide range of along-track view angles makes it feasible to more accurately evaluate the surface contribution to the TOA radiances and hence retrieve aerosols over both ocean and land surfaces, including bright desert and sunglint regions (Diner et al., 1998; Martonchik et al., 1998a; 2002; Kahn et al., 2005a). MISR AODs are within 20% or ±0.05 of coincident AERONET measurements (Kahn et al., 2005a; Abdou et al., 2005). The MISR multi-angle data also sample scattering angles ranging from about 60° to 160° in midlatitudes, yielding information about particle size (Kahn et al., 1998; 2001; 2005a; Chen et al., 2008) and shape (Kalashnikova and Kahn, 2006). The aggregate of aerosol microphysical properties can be used to assess aerosol airmass type, a more robust characterization of MISR-retrieved particle property information than individual attributes. MISR also retrieves plume height in the vicinity of wildfire, volcano, and mineral dust aerosol sources, where the plumes have discernable spatial contrast in the multi-angle imagery (Kahn et al., 2007a). Figure 2.2 is an example that illustrates MISR's ability to characterize the load, optical properties, and stereo height of near-source fire plumes.

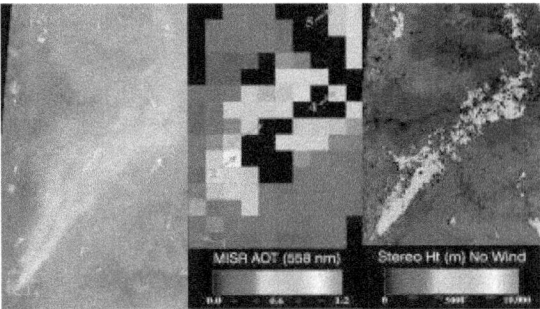

Figure 2.2. Oregon fire on September 4, 2003 as observed by MISR: (a) MISR nadir view of the fire plume, with five patch locations numbered and wind-vectors superposed in yellow; (b) MISR aerosol optical depth at 558 nm; and (c) MISR stereo height without wind correction for the same region (taken from Kahn et al., 2007a).

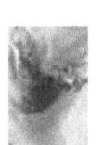

Box 2.3: POLarization and Directionality of the Earth's Reflectance

POLDER is a unique aerosol sensor that consists of wide field-of-view imaging spectro-radiometer capable of measuring multi-spectral, multi-directional, and polarized radiances (Deuzé et al., 2001). The observed radiances can be exploited to better separate the atmospheric contribution from the surface contribution over both land and ocean. POLDER -1 and -2 flew onboard the ADEOS (Advanced Earth Observing Satellite) from November 1996 to June 1997 and April to October of 2003, respectively. A similar POLDER instrument flies on the PARASOL satellite that was launched in December 2004.

Figure 2.3 shows global horizontal patterns of AOD and Ångström exponent over the oceans derived from the POLDER instrument for June 1997. The oceanic AOD map (Figure 2.3.a) reveals near-coastal plumes of high AOD, which decrease with distance from the coast. This pattern arises from aerosol emissions from the continents, followed by atmospheric dispersion, transformation, and removal in the downwind direction. In large-scale flow fields, such as the trade winds, these continental plumes persist over several thousand kilometers. The Ångström exponent shown in Figure 2.3.b exhibits a very different pattern from that of the aerosol optical depth; specifically, it exhibits high values downwind of industrialized regions and regions of biomass burning, indicative of small particles arising from direct emissions from combustion sources and/or gas-to-particle conversion, and low values associated with large particles in plumes of soil dust from deserts and in sea salt aerosols.

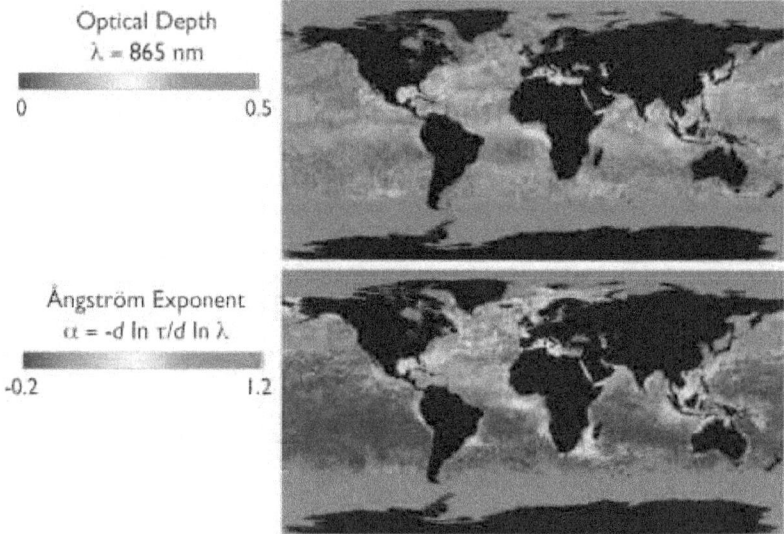

Figure 2.3. Global maps at 18 km resolution showing monthly average (a) AOD at 865 nm and (b) Ångström exponent of AOD over water surfaces only for June, 1997, derived from radiance measurements by the POLDER. Reproduced with permission of Laboratoire d'Optique Atmospherique (LOA), Lille, FR; Laboratoire des Sciences du Climat et de l'Environnement (LSCE), Gif sur Yvette, FR; Centre National d'etudes Spatiales (CNES), Toulouse, FR; and NAtional Space Development Agency (NASDA), Japan.

as-yet-unspecified new technology. Space-based lidars are for the first time providing global constraints on aerosol vertical distribution, and multi-angle imaging is supplementing this with maps of plume injection height in aerosol source regions. Major advances have also been made during the past decade in distinguishing aerosol types from space, and the data are now useful for validating aerosol transport model simulations of aerosol air mass type distributions and transports, particularly over dark water. But particle size, shape, and especially SSA information has large uncertainty; improvements will be needed to better distinguish anthropogenic from natural aerosols using space-based retrievals. The particle microphysical property detail required to assess aerosol radiative forcing will come largely from targeted *in situ* and surface remote sensing measurements, at least for the near-future, although estimates of measurement-based aerosol RF can be made from judicious use of the satellite data with relaxed requirements for characterizing aerosol microphysical properties.

2.2.2. FOCUSED FIELD CAMPAIGNS

Over the past two decades, numerous focused field campaigns have examined the physical, chemical, and optical properties and radiative forcing of aerosols in a variety of aerosol regimes around the world, as listed in Table 2.2. These campaigns, which have been designed with aerosol characterization as the main goal or as one of the major themes in more interdisciplinary studies, were conducted mainly over or downwind of known continental aerosol source regions, but in some instances in low-aerosol regimes, for contrast. During each of these comprehensive campaigns, aerosols were studied in great detail, using combinations of *in situ* and remote sensing observations of physical and chemical properties from various platforms (e.g., aircraft, ships, satellites, and ground-based stations) and numerical modeling. In spite of their relatively short duration, these field studies have acquired comprehensive data sets of regional aerosol properties that have been used to understand the properties and evolution of aerosols within the atmosphere and to improve the climatology of aerosol microphysical properties used in satellite retrieval algorithms and CTMs.

2.2.3. GROUND-BASED IN SITU MEASUREMENT NETWORKS

Major US-operated surface *in situ* and remote sensing networks for tropospheric aerosol characterization and climate forcing research are listed in Table 2.3. These surface *in situ* stations provide information about long-term changes and trends in aerosol concentrations and properties, the influence of regional sources on aerosol properties, climatologies of aerosol radiative properties, and data for testing models (e.g., Quinn et al., 2000; Quinn et al., 2002; Delene and Ogren, 2002; Sheridan and Ogren, 1999; Fiebig and Ogren, 2006; Bates et al., 2006; Quinn et al., 2007) and satellite aerosol retrievals. The NOAA Earth System Research Laboratory (ESRL) aerosol monitoring network consists of baseline, regional, and mobile stations. These near-surface measurements include submicrometer and sub-10 micrometer scattering and absorption coefficients from which the extinction coefficient and single-scattering albedo can be derived. Additional measurements include particle concentration and, at selected sites, CCN concentration, the hygroscopic growth factor, and chemical composition.

Several of the stations, which are located across North America and world-wide, are in regions where recent focused field campaigns have been conducted. The measurement protocols at the stations are similar to those used during the field campaigns. Hence, the station data are directly comparable to the field campaign data so that they provide a longer-term measure of mean aerosol properties and their variability, as well as a context for the shorter-duration measurements of the field campaigns.

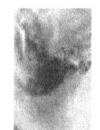

Box 2.4: Clouds and the Earth's Radiant Energy System

CERES measures broadband solar and terrestrial radiances at three channels with a large footprint (e.g., 20 km for CERES/Terra) (Wielicki et al., 1996). It is collocated with MODIS and MISR aboard Terra and with MODIS on Aqua. The observed radiances are converted to TOA irradiances or fluxes using the Angular Distribution Models (ADMs) that are functions of viewing angle, sun angle, and scene type (Loeb and Kato, 2002; Zhang et al., 2005a; Loeb et al., 2005). Such estimates of TOA solar flux in clear-sky conditions can be compared to the expected flux for an aerosol-free atmosphere, in conjunction with measurements of aerosol optical depth from other sensors (e.g., MODIS and MISR) to derive the aerosol direct radiative forcing (Loeb and Manalo-Smith, 2005; Zhang and Christopher, 2003; Zhang et al., 2005b; Christopher et al., 2006; Patadia et al., 2008). The derived instantaneous value is then scaled to obtain a daily average. A direct use of the coarse spatial resolution CERES measurements would exclude aerosol distributions in partly cloudy CERES scenes. Several approaches that incorporate coincident, high spatial and spectral resolution measurements (e.g., MODIS) have been employed to overcome this limitation (Loeb and Manalo-Smith, 2005; Zhang et al., 2005b).

The Interagency Monitoring of Protected Visual Environment (IMPROVE), which is operated by the National Park Service Air Resources Division, has stations across the US located within national parks (Malm et al., 1994). Although the primary focus of the network is air pollution, the measurements are also relevant to climate forcing research. Measurements include fine and coarse mode (PM2.5 and PM10) aerosol mass concentration; concentrations of elements, sulfate, nitrate, organic carbon, and elemental carbon; and scattering coefficients.

In addition, to these US-operated networks, there are other national and international surface networks that provide measurements of aerosol properties including, but not limited to, the World Meteorological Organization (WMO) Global Atmospheric Watch (GAW) network (http://www.wmo.int/pages/prog/arep/gaw/

Box 2.5: Active Remote Sensing of Aerosols

Following the success of a demonstration of lidar system aboard the U.S. Space Shuttle mission in 1994, i.e., Lidar In-space Technology Experiment (LITE) (Winker et al., 1996), the Geoscience Laser Altimeter System (GLAS) was launched in early 2003 to become the first polar orbiting satellite lidar. It provides global aerosol and cloud profiling for a one-month period out of every three-to-six months. It has been demonstrated that GLAS is capable of detecting and discriminating multiple layer clouds, atmospheric boundary layer aerosols, and elevated aerosol layers (e.g., Spinhirne et al., 2005). The Cloud-Aerosol Lidar and Infrared Pathfinder Satellite Observations (CALIPSO), launched on April 28, 2006, is carrying a lidar instrument (Cloud and Aerosol Lidar with Orthogonal Polarization - CALIOP) that has been collecting profiles of the attenuated backscatter at visible and near-infrared wavelengths along with polarized backscatter in the visible channel (Winker et al., 2003). CALIOP measurements have been used to derive the above-cloud fraction of aerosol extinction optical depth (Chand et al., 2008), one of the important factors determining aerosol direct radiative forcing in cloudy conditions. Figure 2.4 shows an event of trans-Atlantic transport of Saharan dust captured by CALIPSO. Flying in formation with the Aqua, AURA, POLDER, and CloudSat satellites, the vertically resolved information is expected to greatly improve passive aerosol and cloud retrievals as well as allow the retrieval of vertical distributions of aerosol extinction, fine- and coarse-mode separately (Kaufman et al., 2003; Leon et al., 2003; Huneeus and Boucher, 2007).

Figure 2.4. A dust event that originated in the Sahara desert on 17 August 2007 and was transported to the Gulf of Mexico. Red lines represent back trajectories indicating the transport track of the dust event. Vertical images are 532 nm attenuated backscatter coefficients measured by CALIOP when passing over the dust transport track. The letter "D" designates the dust layer, and "S" represents smoke layers from biomass burning in Africa (17–19 August) and South America (22 August). The track of the high-spectral-resolution-lidar (HSRL) measurement is indicated by the white line superimposed on the 28 August CALIPSO image. The HSRL track is coincident with the track of the 28 August CALIPSO measurement off the coast of Texas between 28.75°N and 29.08°N (taken from Liu et al., 2008).

monitoring.html), the European Monitoring and Evaluation Programme (EMEP) (http://www.emep.int/), the Canadian Air and Precipitation Monitoring Network (CAPMoN) (http://www.msc-smc.ec.gc.ca/capmon/index_e.cfm), and the Acid Deposition Monitoring Network in East Asia (EANET) (http://www.eanet.cc/eanet.html).

2.2.4. IN SITU AEROSOL PROFILING PROGRAMS

In addition to long-term ground based measurements, regular long-term aircraft *in situ* measurements recently have been implemented at several locations. These programs provide a statistically significant data set of the vertical distribution of aerosol properties to determine

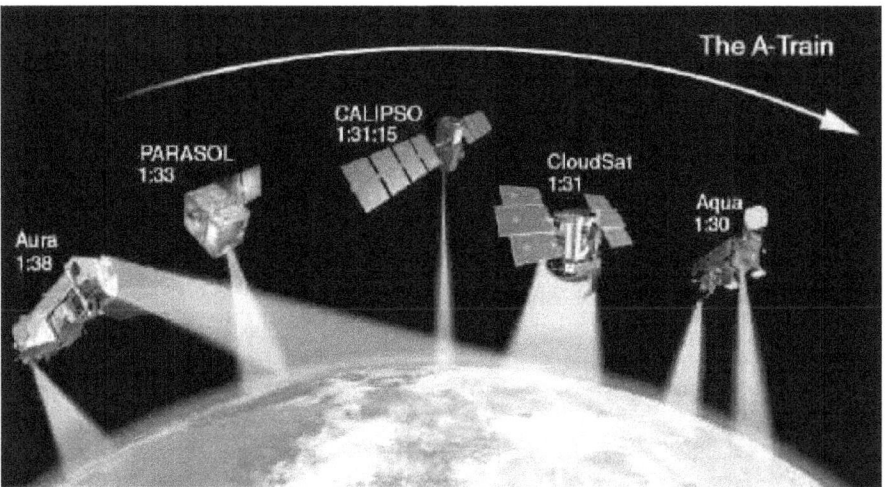

Figure 2.5. A constellation of five spacecraft that overfly the Equator at about 1:30 PM, the so-called A-Train, carries sensors having complementary capabilities, offering unprecedented opportunities to study aerosols from space in multiple dimensions.

spatial and temporal variability through the vertical column and the influence of regional sources on that variability. In addition, the measurements provide data for satellite and model validation. As part of its long-term ground measurements, NOAA has conducted regular flights over Bondville, Illinois since 2006. Measurements include light scattering and absorption coefficients, the relative humidity dependence of light scattering, aerosol number concentration and size distribution, and chemical composition. The same measurements with the exception of number concentration, size distribution, and chemical composition were made by NOAA during regular overflights of DOE ARM's Southern Great Plains (SGP) site from 2000 to 2007 (Andrews et al., 2004) (http://www.esrl. noaa.gov/gmd/aero/net/index.html).

In summary of sections 2.2.2, 2.2.3, and 2.2.4, *in situ* measurements of aerosol properties have greatly expanded over the past two decades as evidenced by the number of focused field campaigns in or downwind of aerosol source regions all over the globe, the continuation of existing and implementation of new sampling networks worldwide, and the implementation of regular aerosol profiling measurements from fixed locations. In addition, *in situ* measurement capabilities have undergone major advancements during this same time period. These advancements include the ability to measure aerosol chemical composition as a function of size at a time resolution of seconds to minutes (e.g., Jayne et al., 2000), the devel-

opment of instruments able to measure aerosol absorption and extinction coefficients at high sensitivity and time resolution and as a function of relative humidity (e.g., Baynard et al., 2007; Lack et al., 2006), and the deployment of these instruments across the globe on ships, at ground-based sites, and on aircraft. However, further advances are needed to make this newly developed instrumentation more affordable and turn-key so that it can be deployed more widely to characterize aerosol properties at a variety of sites world-wide.

2.2.5. GROUND-BASED REMOTE SENSING MEASUREMENT NETWORKS

The Aerosol Robotic Network (AERONET) program is a federated ground-based remote sensing network of well-calibrated sun photometers and radiometers (http://aeronet.gsfc. nasa.gov).

AERONET includes about 200 sites around the world, covering all major tropospheric aerosol regimes (Holben et al., 1998; 2001), as illustrated in Figure 2.6. Spectral measurements of sun and sky radiance are calibrated and screened for cloud-free conditions (Smirnov et al., 2000). AERONET stations provide direct, calibrated measurements of spectral AOD (normally at wavelengths of 440, 670, 870, and 1020 nm) with an accuracy of ±0.015 (Eck et al., 1999). In addition, inversion-based retrievals of a variety of effective, column-mean properties have been developed, including aerosol single-scattering albedo, size distributions, fine-mode frac-

In situ measurements of aerosol properties have greatly expanded over the past two decades as evidenced by the number of focused field campaigns, the world-wide sampling networks, and the regular aerosol profiling measurements from fixed locations.

Table 2.2. List of major intensive field experiments that are relevant to aerosol research in a variety of aerosol regimes around the globe conducted in the past two decades (updated from Yu et al., 2006).

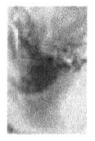

Aerosol Regimes	Intensive Field Experiments			Major References
	Name	**Location**	**Time Period**	
Anthropogenic aerosol and boreal forest from North America and West Europe	TARFOX	North Atlantic	July 1996	Russell et al., 1999
	NEAQS	North Atlantic	July-August 2002	Quinn and Bates, 2003
	SCAR-A	North America	1993	Remer et al., 1997
	CLAMS	East Coast of U.S.	July-August 2001	Smith et al., 2005
	INTEX-NA, ICARTT	North America	Summer 2004	Fehsenfeld et al., 2006
	DOE AIOP	northern Oklahoma	May 2003	Ferrare et al., 2006
	MILAGRO	Mexico city, Mexico	March 2006	Molina et al., 2008
	TexAQS/ GoMACCS	Texas and Gulf of Mexico	August-September 2006	Jiang et al., 2008; Lu et al., 2008
	ARCTAS	North-central Alaska to Greenland (Arctic haze)	March-April 2008	http://www.espo.nasa.gov/arctas/
	ARCTAS	Northern Canada (smoke)	June-July 2008	
	ACE-2	North Atlantic	June-July 1997	Raes et al., 2000
	MINOS	Mediterranean region	July-August 2001	Lelieveld et al., 2002
	LACE98	Lindberg, Germany	July-August 1998	Ansmann et al., 2002
	Aerosols99	Atlantic	January-February 1999	Bates et al., 2001
Brown Haze in South Asia	INDOEX	Indian subcontinent and Indian Ocean	January-April 1998 and 1999	Ramanathan et al., 2001b
	ABC	South and East Asia	ongoing	Ramanathan and Crutzen, 2003
Anthropogenic aerosol and desert dust mixture from East Asia	EAST-AIRE	China	March-April 2005	Li et al., 2007
	INTEX-B	northeastern Pacific	April 2006	Singh et al., 2008
	ACE-Asia	East Asia and Northwest Pacific	April 2001	Huebert et al., 2003; Seinfeld et al., 2004
	TRACE-P		March-April 2001	Jacob et al., 2003
	PEM-West A & B	Western Pacific off East Asia	September-October 1991 February-March 1994	Hoell et al., 1996; 1997
Biomass burning smoke in the tropics	BASE-A	Brazil	1989	Kaufman et al., 1992
	SCAR-B	Brazil	August-September 1995	Kaufman et al., 1998
	LBA-SMOCC	Amazon basin	September-November 2002	Andreae et al., 2004
	SAFARI2000	South Africa and South Atlantic	August -September 2000	King et al., 2003
	SAFARI92		September-October 1992	Lindesay et al., 1996
	TRACE-A	South Atlantic	September-October 1992	Fishman et al., 1996
	DABEX	West Africa	January-February 2006	Haywood et al., 2008
Mineral dusts from North Africa and Arabian Peninsula	SAMUM	Southern Morocco	May-June 2006	Heintzenberg et al., 2009
	SHADE	West coast of North Africa	September 2000	Tanré et al., 2003
	PRIDE	Puerto Rico	June-July 2000	Reid et al., 2003
	UAE[2]	Arabian Peninsula	August-September 2004	Reid et al., 2008
Remote Oceanic Aerosol	ACE-1	Southern Oceans	December 1995	Bates et al., 1998; Quinn and Coffman, 1998

Table 2.3. Summary of major US surface *in situ* and remote sensing networks for the tropospheric aerosol characterization and radiative forcing research. All the reported quantities are column-integrated or column-effective, except as indicated.

Surface Network		Measured/derived parameters				Spatial coverage	Temporal coverage
		Loading	Size, shape	Absorption	Chemistry		
In Situ	NOAA ESRL aerosol monitoring (http://www.esrl.noaa.gov/gmd/aero/)	near-surface extinction coefficient, optical depth, CN/CCN number concentrations	Angstrom exponent, hemispheric backscatter fraction, asymmetry factor, hygroscopic growth	single-scattering albedo, absorption coefficient	chemical composition in selected sites and periods	5 baseline stations, several regional stations, aircraft and mobile platforms	1976 onward
	NPS/EPA IMPROVE (http://vista.cira.colostate.edu/improve/)	near-surface mass concentrations and derived extinction coefficients by species	fine and coarse separately	single-scattering albedo, absorption coefficient	ions, ammonium sulfate, ammonium nitrate, organics, elemental carbon, fine soil	156 national parks and wilderness areas in the U.S.	1988 onward
Remote Sensing	NASA AERONET (http://aeronet.gsfc.nasa.gov)	optical depth	fine-mode fraction, Angstrom exponents, asymmetry factor, phase function, non-spherical fraction	single-scattering albedo, absorption optical depth, refractive indices	N/A	~200 sites over global land and islands	1993 onward
	DOE ARM (http://www.arm.gov)					6 sites and 1 mobile facility in N. America, Europe, and Asia	1989 onward
	NOAA SURFRAD (http://www.srrb.noaa.gov/surfrad/)		N/A	N/A	N/A	7 sites in the U.S.	1995 onward
	AERONET- MAN (http://aeronet.gsfc.nasa.gov/maritime_aerosol_network.html)					global ocean	2004-present (periodically)
	NASA MPLNET (http://mplnet.gsfc.nasa.gov/)	vertical profiles of backscatter/extinction coefficient	N/A	N/A	N/A	~30 sites in major continents, usually collocated with AERONET and ARM sites	2000 onward

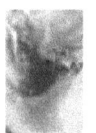

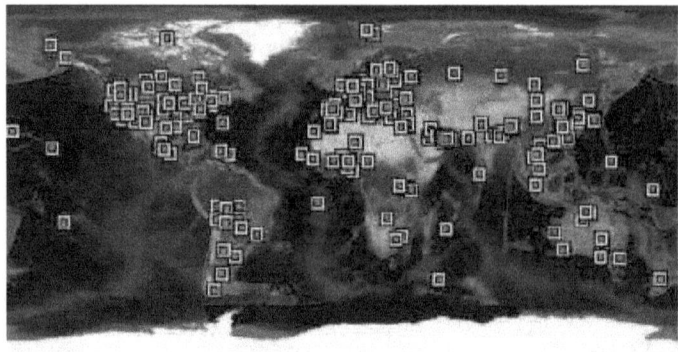

Figure 2.6. Geographical coverage of active AERONET sites in 2006.

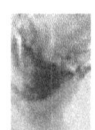

AERONET includes about 200 sites around the world, covering all major tropospheric aerosol regimes. AERONET stations provide direct, calibrated measurements of spectral AOD with an accuracy of ±0.015.

tion, degree of non-sphericity, phase function, and asymmetry factor (Dubovik et al., 2000; Dubovik and King, 2000; Dubovik et al., 2002; O'Neill, et al., 2004). The SSA can be retrieved with an accuracy of ±0.03, but only for AOD >0.4 (Dubovik et al., 2002), which precludes much of the planet. These retrieved parameters have been validated or are undergoing validation by comparison to *in situ* measurements (e.g., Haywood et al., 2003; Magi et al., 2005; Leahy et al., 2007).

Recent developments associated with AERONET algorithms and data products include:
- simultaneous retrieval of aerosol and surface properties using combined AERONET and satellite measurements (Sinyuk et al., 2007) with surface reflectance taken into account (which significantly improves AERONET SSA retrieval accuracy) (Eck et al., 2008);
- the addition of ocean color and high frequency solar flux measurements; and
- the establishment of the Maritime Aerosol Network (MAN) component to monitor aerosols over the World oceans from ships-of-opportunity (Smirnov et al., 2006).

Because of consistent calibration, cloud-screening, and retrieval methods, uniformly acquired and processed data are available from all stations, some of which have operated for over 10 years. These data constitute a high-quality, ground-based aerosol climatology and, as such, have been widely used for aerosol process studies as well as for evaluation and validation of model simulation and satellite remote sensing applications (e.g., Chin et al., 2002; Yu et al., 2003, 2006; Remer et al., 2005; Kahn et al., 2005a). In addition, AERONET retrievals of aerosol size distribution and refractive indices have been used in algorithm development for

satellite sensors (Remer et al., 2005; Levy et al., 2007a). A set of aerosol optical properties provided by AERONET has been used to calculate the aerosol direct radiative forcing (Procopio et al., 2004; Zhou et al., 2005), which can be used to evaluate both satellite remote sensing measurements and model simulations.

AERONET measurements are complemented by other ground-based aerosol networks having less geographical or temporal coverage, such as the Atmospheric Radiation Measurement (ARM) network (Ackerman and Stokes, 2003), NOAA's national surface radiation budget network (SURFRAD) (Augustine et al., 2008) and other networks with multifilter rotating shadowband radiometer (MFRSR) (Harrison et al., 1994; Michalsky et al., 2001), and several lidar networks including
- NASA Micro Pulse Lidar Network (MPL-NET) (Welton et al., 2001; 2002);
- Regional East Atmospheric Lidar Mesonet (REALM) in North America (Hoff et al., 2002; 2004);
- European Aerosol Research Lidar Network (EARLINET) (Matthias et al., 2004); and
- Asian Dust Network (AD-Net) (e.g., Murayama et al., 2001).

Obtaining accurate aerosol extinction profile observations is pivotal to improving aerosol radiative forcing and atmospheric response calculations. The values derived from these lidar networks with state-of-the-art techniques (Schmid et al., 2006) are helping to fill this need.

2.2.6. SYNERGY OF MEASUREMENTS AND MODEL SIMULATIONS

Individual approaches discussed above have their own strengths and limitations, and are usually complementary. None of these approaches alone is adequate to characterize large spatial and temporal variations of aerosol physical and chemical properties and to address complex aerosol-climate interactions. The best strategy for characterizing aerosols and estimating their radiative forcing is to integrate measurements from different satellite sensors with complementary capabilities from *in situ* and surface-based measurements. Similarly, while models are essential tools for estimating regional and global distributions and radiative forcing of aerosols at present as well as in the past and the future, observations are required to provide constraints and validation of the models. In the following, several synergistic approaches to studying aerosols and their radiative forcing are discussed.

Closure experiments: During intensive field studies, multiple platforms and instruments are deployed to sample regional aerosol properties through a well-coordinated experimental design. Often, several independent methods are used to measure or derive a single aerosol property or radiative forcing. This combination of methods can be used to identify inconsistencies in the methods and to quantify uncertainties in measured, derived, and calculated aerosol properties and radiative forcings. This approach, often referred to as a closure experiment, has been widely employed on both individual measurement platforms (local closure) and in studies involving vertical measurements through the atmospheric column by one or more platforms (column closure) (Quinn et al., 1996; Russell et al., 1997).

Past closure studies have revealed that the best agreement between methods occurs for submicrometer, spherical particles such that different measures of aerosol optical properties and optical depth agree within 10 to 15% and often better (e.g., Clarke et al., 1996; Collins et al., 2000; Schmid et al., 2000; Quinn et al., 2004). Larger particle sizes (e.g., sea salt and dust) present inlet collection efficiency issues and non-spherical particles (e.g., dust) lead to differences in instrumental responses. In these cases, differences between methods for determining aerosol optical depth can be as great as 35% (e.g., Wang et al., 2003; Doherty et al., 2005). Closure studies on aerosol clear-sky DRF reveal uncertainties of about 25% for sulfate/carbonaceous aerosol and 60% for dust-containing aerosol (Bates et al., 2006). Future closure studies could integrate surface- and satellite-based radiometric measurements of AOD with *in situ* optical, microphysical, and aircraft radiometric measurements for a wide range of situations. There is also a need to maintain consistency in comparing results and expressing uncertainties (Bates et al., 2006).

Constraining models with *in situ* measurements: *In situ* measurements of aerosol chemical, microphysical, and optical properties with known accuracy, based in part on closure studies, can be used to constrain regional

CTM simulations of aerosol direct forcing, as described by Bates et al. (2006). A key step in the approach is assigning empirically derived optical properties to the individual chemical components generated by the CTM for use in a Radiative Transfer Model (RTM). Specifically, regional data from focused, short-duration field programs can be segregated according to aerosol type (sea salt, dust, or sulfate/carbonaceous) based on measured chemical composition and particle size. Corresponding measured optical properties can be carried along in the sorting process so that they, too, are segregated by aerosol type. The empirically derived aerosol properties for individual aerosol types, including mass scattering efficiency, single-scattering albedo, and asymmetry factor, and their dependences on relative humidity, can be used in place of assumed values in CTMs.

Short-term, focused measurements of aerosol properties (e.g., aerosol concentration and AOD) also can be used to evaluate CTM parameterizations on a regional basis, to suggest improvements to such uncertain model parameters, such as emission factors and scavenging coefficients (e.g., Koch et al., 2007). Improvements in these parameterizations using observations yield increasing confidence in simulations covering regions and periods where and when measurements are not available. To evaluate the aerosol properties generated by CTMs on broader scales in space and time, satellite observations and long-term *in situ* measurements are required.

Improving model simulations with satellite measurements: Global measurements of aerosols from satellites (mainly AOD) with well-defined accuracies offer an opportunity to evaluate model simulations at large spatial and temporal scales. The satellite measurements can also be used to constrain aerosol model simulations and hence the assessment of aerosol DRF through data assimilation or objective analysis process (e.g., Collins et al., 2001; Yu et al., 2003; 2004, 2006; Liu et al., 2005; Zhang et al., 2008). Both satellite retrievals and model simulations have uncertainties. The goal of data integration is to minimize the discrepancies between them, and to form an optimal estimate of aerosol distributions by combining them, typically with weights inversely proportional to the square of the errors of individual descriptions. Such integration can fill gaps in satellite retrievals and

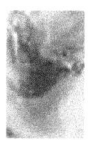

The best strategy for characterizing aerosols and estimating their radiative forcing is to integrate measurements from different satellite sensors with in situ and surface based measurements. Observations are required to provide constraints and validation of the models.

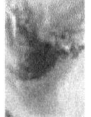

generate global distributions of aerosols that are consistent with ground-based measurements (Collins et al., 2001; Yu et al., 2003, 2006; Liu et al., 2005). Recent efforts have also focused on retrieving global sources of aerosol from satellite observations using inverse modeling, which may be valuable for reducing large aerosol simulation uncertainties (Dubovik et al., 2007). Model refinements guided by model evaluation and integration practices with satellite retrievals can then be used to improve aerosol simulations of the pre- and post-satellite eras.

Current measurement-based understanding of aerosol characterization and radiative forcing is assessed in Section 2.3 through intercomparisons of a variety of measurement-based estimates and model simulations published in literature. This is followed by a detailed discussion of major outstanding issues in section 2.4.

2.3. Assessments of Aerosol Characterization and Climate Forcing

This section focuses on the assessment of measurement-based aerosol characterization and its use in improving estimates of the direct

radiative forcing on regional and global scales. *In situ* measurements provide highly accurate aerosol chemical, microphysical, and optical properties on a regional basis and for the particular time period of a given field campaign. Remote sensing from satellites and ground-based networks provide spatial and temporal coverage that intensive field campaigns lack. Both *in situ* measurements and remote sensing have been used to determine key parameters for estimating aerosol direct radiative forcing including aerosol single scattering albedo, asymmetry factor, optical depth Remote sensing has also been providing simultaneous measurements of aerosol optical depth and radiative fluxes that can be combined to derive aerosol direct radiative forcing at the TOA with relaxed requirement for characterizing aerosol properties. Progress in using both satellite and surface-based measurements to study aerosol-cloud interactions and aerosol indirect forcing is also discussed.

2.3.1. The Use of Measured Aerosol Properties to Improve Models

The wide variety of aerosol data sets from intensive field campaigns provides a rigorous "testbed" for model simulations of aerosol properties and distributions and estimates of DRF. As described in Section 2.2.6, *in situ* measurements can be used to constrain regional CTM simulations of aerosol properties, DRF, anthropogenic component of DRF, and to evaluate CTM parameterizations. In addition, *in situ* measurements can be used to develop simplifying parameterizations for use by CTMs.

Several factors contribute to the uncertainty of CTM calculations of size-distributed aerosol composition including emissions, aerosol removal by wet deposition, processes involved in the formation of secondary aerosols and the chemical and microphysical evolution of aerosols, vertical transport, and meteorological fields including the timing and amount of precipitation, formation of clouds, and relative humidity. *In situ* measurements made during focused field campaigns provide a point of comparison for the CTM-generated aerosol distributions at the surface and at discrete points above the surface. Such comparisons are essential for identifying areas where the models need improvement.

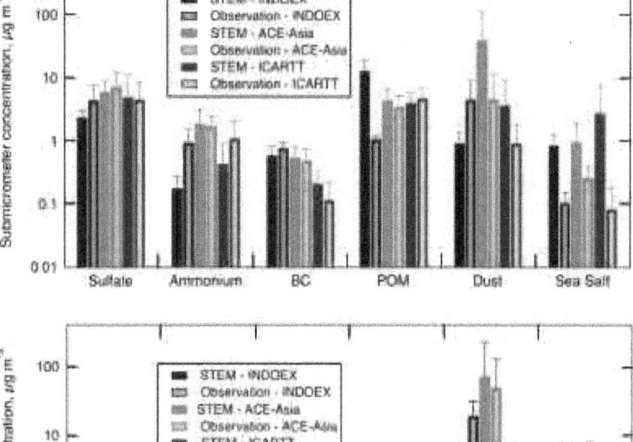

Figure 2.7. Comparison of the mean concentration (µg m-3) and standard deviation of the modeled (STEM) aerosol chemical components with shipboard measurements during INDOEX, ACE-Asia, and ICARTT. After Bates et al. (2006).

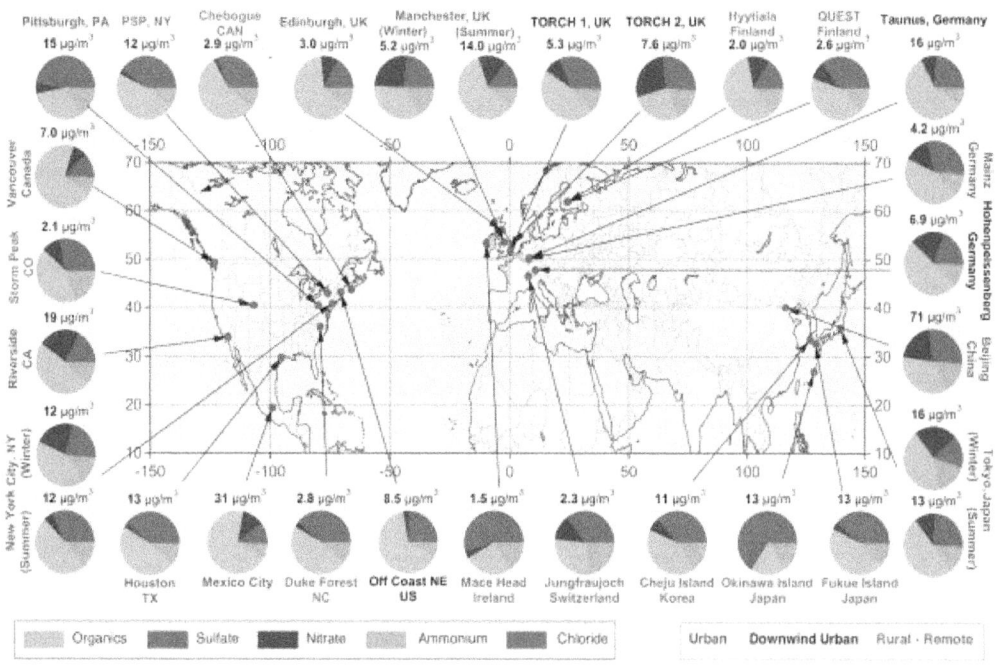

Figure 2.8. Location of aerosol chemical composition measurements with aerosol mass spectrometers. Colors for the labels indicate the type of sampling location: urban areas (blue), <100 miles downwind of major cites (black), and rural/remote areas >100 miles downwind (pink). Pie charts show the average mass concentration and chemical composition: organics (green), sulfate (red), nitrate (blue), ammonium (orange), and chloride (purple), of non-refractory PM1. Adapted from Zhang et al. (2007).

Figure 2.7 shows a comparison of submicrometer and supermicrometer aerosol chemical components measured during INDOEX, ACE-Asia, and ICARTT onboard a ship and the same values calculated with the STEM Model (e.g., Carmichael et al., 2002, 2003; Tang et al., 2003, 2004; Bates et al., 2004; Streets et al., 2006b). To permit direct comparison of the measured and modeled values, the model was driven by analyzed meteorological data and sampled at the times and locations of the shipboard measurements every 30 min along the cruise track. The best agreement was found for submicrometer sulfate and BC. The agreement was best for sulfate; this is attributed to greater accuracy in emissions, chemical conversion, and removal for this component. Underestimation of dust and sea salt is most likely due to errors in model-calculated emissions. Large discrepancies between the modeled and measured values occurred for submicrometer particulate organic matter (POM) (INDOEX), and for particles in the supermicrometer size range such as dust (ACE-Asia), and sea salt (all regions). The model underestimated the total mass of the supermicrometer aerosol by about a factor of 3.

POM makes up a large and variable fraction of aerosol mass throughout the anthropogenically influenced northern hemisphere, and yet models have severe problems in properly representing this type of aerosol. Much of this discrepancy follows from the models inability to represent the formation of secondary organic aerosols (SOA) from the precursor volatile organic compounds (VOC). Figure 2.8 shows a summary of the results from aerosol mass spectrometer measurements at 30 sites over North America, Europe, and Asia. Based on aircraft measurements of urban-influenced air over New England, de Gouw et al. (2005) found that POM was highly correlated with secondary anthropogenic gas phase species suggesting that the POM was derived from secondary anthropogenic sources and that the formation took one day or more.

Figure 2.9 shows scatterplots of submicrometer POM versus acetylene (a gas phase primary emitted VOC species) and isopropyl nitrate (a secondary gas phase organic species formed by atmospheric reactions). The increase in submicrometer POM with increasing photochemical age could not be explained by the removal of

Particulate organic matter makes up a large and variable fraction of aerosol mass throughout the northern hemisphere, and yet models have severe problems in properly representing this type of aerosol.

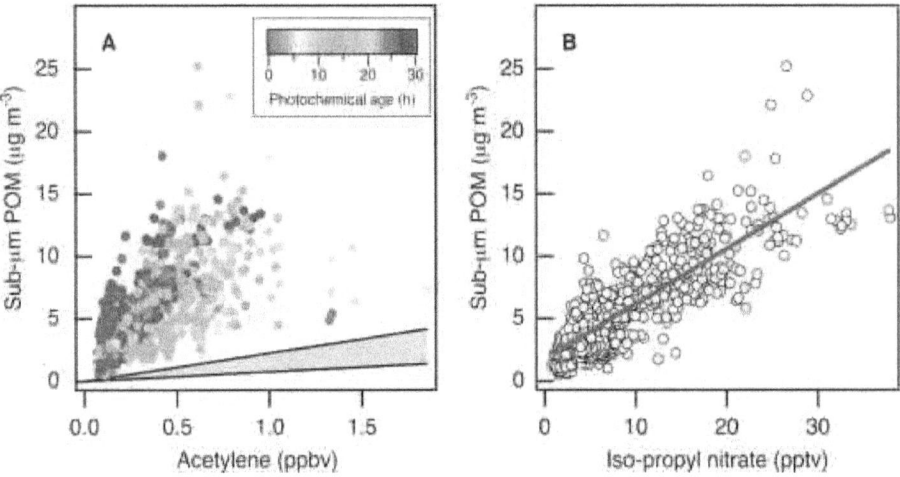

Figure 2.9. Scatterplots of the submicrometer POM measured during NEAQS versus a) acetylene and b) iso-propyl nitrate. The colors of the data points in a) denote the photochemical age as determined by the ratios of compounds of known OH reactivity. The gray area in a) shows the range of ratios between submicrometer POM and acetylene observed by Kirchstetter et al. (1999) in tunnel studies. Adapted from de Guow et al. (2005).

calculate dry mass concentrations of the dominant aerosol species (sulfate, organic carbon, black carbon, sea salt, and dust). *In situ* measurements were used to calculate the corresponding optical properties for each aerosol type for use in a radiative transfer model. Aerosol DRF and its anthropogenic component estimated using the empirically derived and *a priori* optical properties were then compared. The DRF and its anthropogenic component were calculated as the net downward solar flux difference between the model state with aerosol and of the model state with no aerosol. It was found that the constrained optical properties derived from measurements increased the calculated AOD (34 ± 8%), TOA DRF (32 ± 12%), and anthropogenic component of TOA DRF (37 ± 7%) relative to runs using the *a priori* values. These increases were due to larger values of the constrained mass extinction efficiencies relative to the *a priori* values. In addition, differences in AOD due to using the aerosol loadings from MOZART versus those from STEM were much greater than differences resulting from the *a priori* vs. constrained RTM runs.

In situ observations also can be used to generate simplified parameterizations for CTMs and RTMs thereby lending an empirical foundation to uncertain parameters currently in use by models. CTMs generate concentration fields of individual aerosol chemical components that are then used as input to radiative transfer models (RTMs) for the calculation of DRF. Currently, these calculations are performed with a variety of simplifying assumptions concerning the RH dependence of light scattering by the aerosol. Chemical components often are treated as externally mixed each with a unique RH dependence of light scattering. However, both model and measurement studies reveal that POM, internally mixed with

VOC alone, which are its traditionally recognized precursors. This result suggests that other species must have contributed and/or that the mechanism for POM formation is more efficient than assumed by models. Similar results were obtained from the 2006 MILAGRO field campaign conducted in Mexico City (Kleinman et al., 2008), and comparisons of GCM results with several long-term monitoring stations also showed that the model underestimated organic aerosol concentrations (Koch et al., 2007). Recent laboratory work suggests that isoprene may be a major SOA source missing from previous atmospheric models (Kroll et al., 2006; Henze and Seinfeld, 2006), but underestimating sources from certain economic sectors may also play a role (Koch et al., 2007). Models also have difficulty in representing the vertical distribution of organic aerosols, underpredicting their occurrence in the free troposphere (FT) (Heald et al., 2005). While organic aerosol presents models with some of their greatest challenges, even the distribution of well-characterized sulfate aerosol is not always estimated correctly in models (Shindell et al., 2008a).

Comparisons of DRF and its anthropogenic component calculated with assumed optical properties and values constrained by *in situ* measurements can help identify areas of uncertainty in model parameterizations. In a study described by Bates et al. (2006), two different CTMs (MOZART and STEM) were used to

water-soluble salts, can reduce the hygroscopic response of the aerosol, which decreases its water content and ability to scatter light at elevated relative humidity (e.g., Saxena et al., 1995; Carrico et al., 2005). The complexity of the POM composition and its impact on aerosol optical properties requires the development of simplifying parameterizations that allow for the incorporation of information derived from field measurements into calculations of DRF (Quinn et al., 2005). Measurements made during INDOEX, ACE-Asia, and ICARTT revealed a substantial decrease in $f_{\sigma sp}(RH)$ with increasing mass fraction of POM in the accumulation mode. Based on these data, a parameterization was developed that quantitatively describes the relationship between POM mass fraction and $f_{\sigma sp}(RH)$ for accumulation mode sulfate-POM mixtures (Quinn et al., 2005). This simplified parameterization may be used as input to RTMs to derive values of $f_{\sigma sp}(RH)$ based on CTM estimates of the POM mass fraction. Alternatively, the relationship may be used to assess values of $f_{\sigma sp}(RH)$ currently being used in RTMs.

2.3.2. INTERCOMPARISONS OF SATELLITE MEASUREMENTS AND MODEL SIMULATION OF AEROSOL OPTICAL DEPTH

As aerosol DRF is highly dependent on the amount of aerosol present, it is of first-order importance to improve the spatial characterization of AOD on a global scale. This requires an evaluation of the various remote sensing AOD data sets and comparison with model-based AOD estimates. The latter comparison is particularly important if models are to be used in projections of future climate states that would result from assumed future emissions. Both remote sensing and model simulation have uncertainties and satellite-model integration is needed to obtain an optimum description of aerosol distribution.

Figure 2.10 shows an intercomparison of annual average AOD at 550 nm from two recent satellite aerosol sensors (MODIS and MISR), five model simulations (GOCART, GISS, SPRINTARS, LMDZ-LOA, LMDZ-INCA) and three satellite-model integrations (MO_GO, MI_GO, MO_MI_GO). These model-satellite integrations are conducted by using an optimum interpolation approach (Yu et al., 2003) to constrain GOCART simulated AOD with that from MODIS, MISR, or MODIS over ocean and MISR over land, denoted as MO_GO, MI_GO, and MO_MI_GO, respectively. MODIS values of AOD are from Terra Collection 4 retrievals and MISR AOD is based on early post launch retrievals. MODIS and MISR retrievals give a comparable average AOD on the global scale, with MISR greater than MODIS by 0.01~0.02 depending on the season. However, differences between MODIS and MISR are much larger when land and ocean are examined separately: AOD from MODIS is 0.02-0.07 higher over land but 0.03-0.04 lower over ocean than the AOD from MISR. Several major causes for the systematic MODIS-MISR differences have been identified, including instrument calibration and sampling differences, different assumptions about ocean surface boundary conditions made in the individual retrieval algorithms, missing particle property or mixture options in the look-up tables, and cloud screening (Kahn

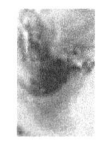

As aerosol direct radiative forcing is highly dependent on the amount of aerosol present, it is of first-order importance to improve the spatial characterization of aerosol optical depth on a global scale.

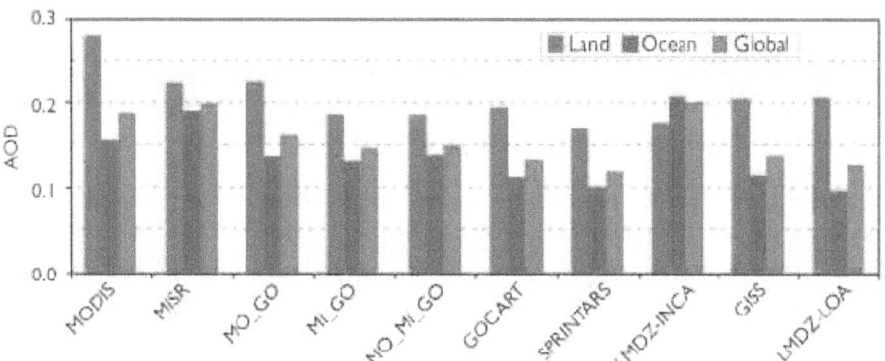

Figure 2.10. Comparison of annual mean aerosol optical depth (AOD) at 550 nm between satellite retrievals (MODIS, MISR), model simulations (GOCART, SPRINTARS, GISS, LMDZ-INCA, LMDZ-LOA), and satellite-model integrations (MO_GO, MI_GO, MO_MI_GO) averaged over land, ocean, and globe (all limited to 60°S-60°N region) (figure generated from Table 6 in Yu et al., 2006).

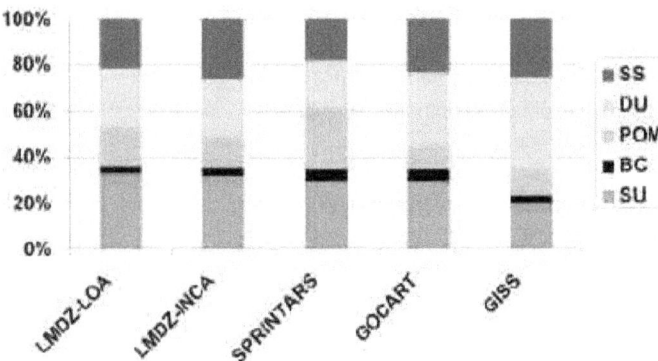
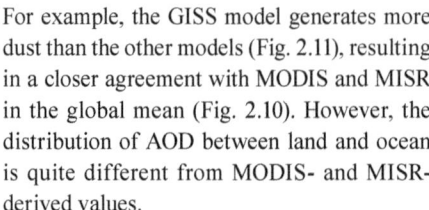

Figure 2.11. Percentage contributions of individual aerosol components (SU – sulfate, BC – black carbon, POM – particulate organic matter, DU – dust, SS – sea salt) to the total aerosol optical depth (at 550 nm) on a global scale simulated by the five models (data taken from Kinne et al., 2006).

et al., 2007b). The MODIS-MISR AOD differences are being reduced by continuous efforts on improving satellite retrieval algorithms and radiance calibration. The new MODIS aerosol retrieval algorithms in Collection 5 have resulted in a reduction of 0.07 for global land mean AOD (Levy et al., 2007b), and improved radiance calibration for MISR removed ~40% of AOD bias over dark water scenes (Kahn et al., 2005b).

The annual and global average AOD from the five models is 0.19±0.02 (mean ± standard deviation) over land and 0.13±0.05 over ocean, respectively. Clearly, the model-based mean AOD is smaller than both MODIS- and MISR-derived values (except the GISS model). A similar conclusion has been drawn from more extensive comparisons involving more models and satellites (Kinne et al., 2006). On regional scales, satellite-model differences are much larger. These differences could be attributed in part to cloud contamination (Kaufman et al., 2005b; Zhang et al., 2005c) and 3D cloud effects in satellite retrievals (Kaufman et al., 2005b; Wen et al., 2006) or to models missing important aerosol sources/sinks or physical processes (Koren et al., 2007b). Integrated satellite-model products are generally in-between the satellite retrievals and the model simulations, and agree better with AERONET measurements (e.g., Yu et al., 2003).

As in comparisons between models and *in situ* measurements (Bates et al., 2006), there appears to be a relationship between uncertainties in the representation of dust in models and the uncertainty in AOD, and its global distribution.

For example, the GISS model generates more dust than the other models (Fig. 2.11), resulting in a closer agreement with MODIS and MISR in the global mean (Fig. 2.10). However, the distribution of AOD between land and ocean is quite different from MODIS- and MISR-derived values.

Figure 2.11 shows larger model differences in the simulated percentage contributions of individual components to the total aerosol optical depth on a global scale, and hence in the simulated aerosol single-scattering properties (e.g., single-scattering albedo, and phase function), as documented in Kinne et al. (2006). This, combined with the differences in aerosol loading (as characterized by AOD) determines the model diversity in simulated aerosol direct radiative forcing, as discussed later. However, current satellite remote sensing capability is not sufficient to constrain model simulations of aerosol components.

2.3.3. Satellite Based Estimates of Aerosol Direct Radiative Forcing

Table 2.4 summarizes approaches to estimating the aerosol direct radiative forcing, including a brief description of methods, identifies major sources of uncertainty, and provides references. These estimates fall into three broad categories, namely (A) satellite-based, (B) satellite-model integrated, and (C) model-based. As satellite aerosol measurements are generally limited to cloud-free conditions, the discussion here focuses on assessments of clear-sky aerosol direct radiative forcing, a net (downwelling minus upwelling) solar flux difference between with aerosol (natural + anthropogenic) and in the absence of aerosol.

Global distributions. Figure 2.12 shows global distributions of aerosol optical depth at 550 nm (left panel) and diurnally averaged clear-sky TOA DRF (right panel) for March-April-May (MAM) based on the different approaches. The DRF at the surface follows the same pattern as that at the TOA but is significantly larger in magnitude because of aerosol absorption. It appears that different approaches agree on large-scale patterns of aerosol optical depth and the direct radiative forcing. In this season, the aerosol impacts in the Northern Hemisphere are much larger than those in the Southern Hemisphere. Dust outbreaks and biomass burning elevate the optical depth to more than 0.3 over

Table 2.4. Summary of approaches to estimating the aerosol direct radiative forcing in three categories: (A) satellite retrievals; (B) satellite-model integrations; and (C) model simulations. (adapted from Yu et al., 2006).

Category	Product	Brief Descriptions	Identified Sources of Uncertainty	Major References
A. Satellite retrievals	MODIS	Using MODIS retrievals of a linked set of AOD, ω_0, and phase function consistently in conjunction with a radiative transfer model (RTM) to calculate TOA fluxes that best match the observed radiances.	Radiance calibration, cloud-aerosol discrimination, instantaneous-to-diurnal scaling, RTM parameterizations	Remer and Kaufman, 2006
	MODIS_A	Splitting MODIS AOD over ocean into mineral dust, sea salt, and biomass-burning and pollution; using AERONET measurements to derive the size distribution and single-scattering albedo for individual components.	Satellite AOD and FMF retrievals, overestimate due to summing up the compositional direct forcing, use of a single AERONET site to characterize a large region	Bellouin et al., 2005
	CERES_A	Using CERES fluxes in combination with standard MODIS aerosol.	Calibration of CERES radiances, large CERES footprint, satellite AOD retrieval, radiance-to-flux conversion (ADM), instantaneous-to-diurnal scaling, narrow-to-broadband conversion	Loeb and Manalo-Smith, 2005; Loeb and Kato, 2002
	CERES_B	Using CERES fluxes in combination with NOAA NESDIS aerosol from MODIS radiances.		
	CERES_C	Using CERES fluxes in combination with MODIS (ocean) and MISR (non-desert land) aerosol with new angular models for aerosols.		Zhang et al., 2005a,b; Zhang and Christopher, 2003; Christopher et al., 2006; Patadia et al., 2008
	POLDER	Using POLDER AOD in combination with prescribed aerosol models (similar to MODIS).	Similar to MODIS	Boucher and Tanré, 2000; Bellouin et al., 2003
B. Satellite-model integrations	MODIS_G	Using GOCART simulations to fill AOD gaps in satellite retrievals.	Propagation of uncertainties associated with both satellite retrievals and model simulations (but the model-satellite integration approach does result in improved AOD quality for MO_GO, and MO_MI_GO)	*Aerosol single-scattering albedo and asymmetry factor are taken from GOCART simulations; *Yu et al., 2003, 2004, 2006
	MISR_G			
	MO_GO	Integration of MODIS and GOCART AOD.		
	MO_MI_GO	Integration of GOCART AOD with retrievals from MODIS (Ocean) and MISR (Land).		
	SeaWiFS	Using SeaWiFS AOD and assumed aerosol models.	Similar to MODIS_G and MISR_G, too weak aerosol absorption	Chou et al., 2002
C. Model simulations	GOCART	Offline RT calculations using monthly average aerosols with a time step of 30 min (without the presence of clouds).	Emissions, parameterizations of a variety of sub-grid aerosol processes (e.g., wet and dry deposition, cloud convection, aqueous-phase oxidation), assumptions on aerosol size, absorption, mixture, and humidification of particles, meteorology fields, not fully evaluated surface albedo schemes, RT parameterizations	Chin et al., 2002; Yu et al., 2004
	SPRINTARS	Online RT calculations every 3 hrs (cloud fraction=0).		Takemura et al., 2002, 2005
	GISS	Online model simulations and weighted by clear-sky fraction.		Koch and Hansen, 2005; Koch et al., 2006
	LMDZ-INCA	Online RT calculations every 2 hrs (cloud fraction = 0).		Balkanski et al., 2007; Schulz et al., 2006; Kinne et al., 2006
	LMDZ-LOA	Online RT calculations every 2 hrs (cloud fraction=0).		Reddy et al., 2005a, b

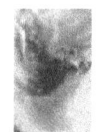

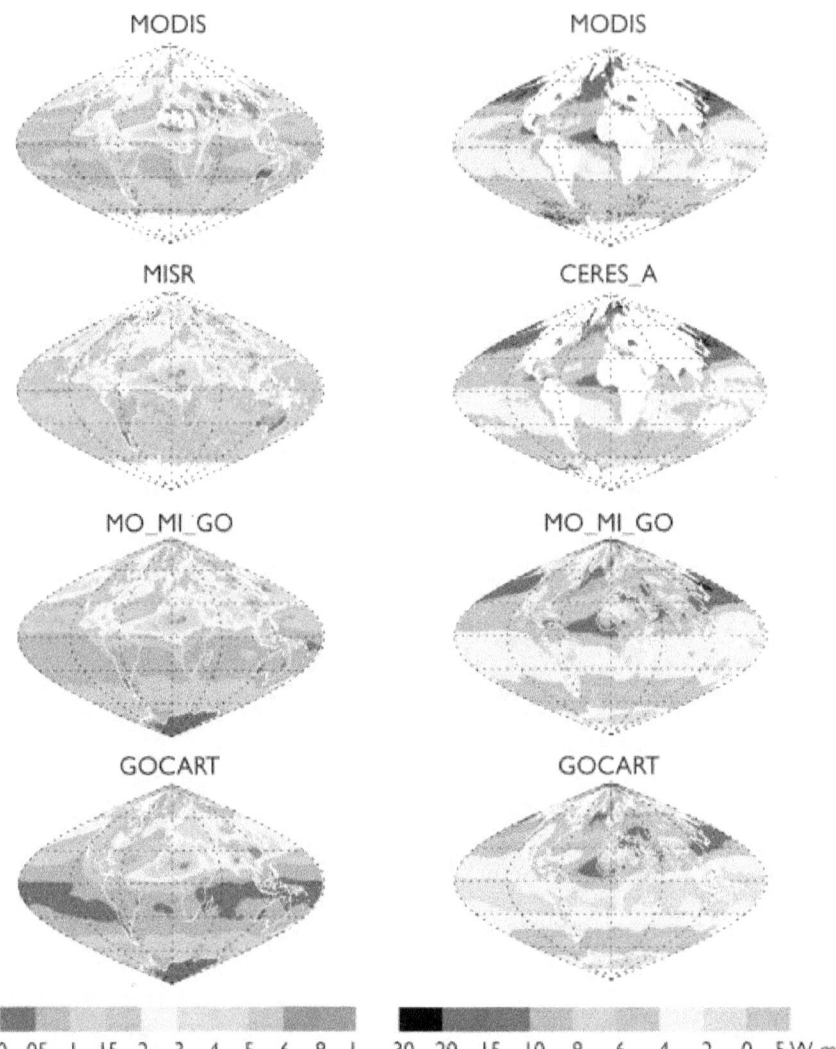

Figure 2.12. Geographical patterns of seasonally (MAM) averaged aerosol optical depth at 550 nm (left panel) and the diurnally averaged clear-sky aerosol direct radiative (solar spectrum) forcing (W m⁻²) at the TOA (right panel) derived from satellite (Terra) retrievals (MODIS, Remer et al., 2005; Remer and Kaufman, 2006; MISR, Kahn et al., 2005a; and CERES_A, Loeb and Manalo-Smith, 2005), GOCART simulations (Chin et al., 2002; Yu et al., 2004), and GOCART-MODIS-MISR integrations (MO_MI_GO, Yu et al., 2006) (taken from Yu et al., 2006).

large parts of North Africa and the tropical Atlantic. In the tropical Atlantic, TOA cooling as large as -10 W m⁻² extends westward to Central America. In eastern China, the optical depth is as high as 0.6-0.8, resulting from the combined effects of industrial activities and biomass burning in the south, and dust outbreaks in the north. The Asian impacts also extend to the North Pacific, producing a TOA cooling of more than -10 W m⁻². Other areas having large aerosol impacts include Western Europe, mid-latitude North Atlantic, and much of South Asia

and the Indian Ocean. Over the "roaring forties" in the Southern Hemisphere, high winds generate a large amount of sea salt. Elevated optical depth, along with high solar zenith angle and hence large backscattering to space, results in a band of TOA cooling of more than -4 W m⁻². However, there is also some question as to whether thin cirrus (e.g., Zhang et al., 2005c) and unaccounted-for whitecaps contribute to the apparent enhancement in AOD retrieved by satellite. Some differences exist between different approaches. For example, the early post-launch MISR retrieved optical depths over the southern hemisphere oceans are higher than MODIS retrievals and GOCART simulations. Over the "roaring forties", the MODIS derived TOA solar flux perturbations are larger than the estimates from other approaches.

Global mean. Figure 2.13 summarizes the measurement- and model-based estimates of clear-sky annual-averaged DRF at both the TOA and surface from 60°S to 60°N. Seasonal DRF values for individual estimates are summarized in Table 2.5 and Table 2.6 for ocean and land, respectively. Mean, median and standard error ε (ε=σ/(n-1)$^{1/2}$), where σ is standard deviation and n is the number of methods) are calculated for measurement- and model-based estimates separately. Note that although the standard deviation or standard error reported here is not a fully rigorous measure of a true experimental uncertainty, it is indicative of the uncertainty because independent approaches with independent sources of errors are used (see Table 2.4; in the modeling community, this is called the "diversity", see Chapter 3).

- **Ocean:** For the TOA DRF, a majority of measurement-based and satellite-model integration-based estimates agree with each other within about 10%. On annual average, the measurement-based estimates give the DRF of -5.5±0.2 W m^{-2} (mean±ε) at the TOA and -8.7±0.7 W m^{-2} at the surface. This suggests that the ocean surface cooling is about 60% larger than the cooling at the TOA. Model simulations give wide ranges of DRF estimates at both the TOA and surface. The ensemble of five models gives the annual average DRF (mean ± ε) of -3.2±0.6 W m^{-2} and -4.9±0.8 W m^{-2} at the TOA and surface, respectively. On average, the surface cooling is about 37% larger than the TOA cooling, smaller than the measurement-based estimate of surface and TOA difference of 60%. However, the 'measurement-based' estimate of *surface* DRF is actually a calculated value, using poorly constrained particle properties.
- **Land:** It remains challenging to use satellite measurements alone for characterizing complex aerosol properties over land surfaces with high accuracy. As such, DRF estimates over land have to rely largely on model simulations and satellite-model inte-

grations. On a global and annual average, the satellite-model integrated approaches derive a mean DRF of -4.9 W m^{-2} at the TOA and -11.9 W m^{-2} at the surface respectively. The surface cooling is more than a factor of 2 larger than the TOA cooling because of aerosol absorption. Note that the TOA DRF of -4.9 W m^{-2} agrees quite well with the most recent satellite-based estimate of -5.1±1.1 W m^{-2} over non-desert land based on coincident measurements of MISR AOD and CERES solar flux (Patadia et al., 2008). For comparisons, an ensemble of five model simulations derives a DRF (mean ± ε) over land of -3.0±0.6 W m^{-2} at the TOA and -7.6±0.9 W m^{-2} at the surface, respectively. Seasonal variations of DRF over land, as derived from both measurements and models, are larger than those over ocean.

The above analyses show that, on a global average, the measurement-based estimates of DRF are 55-80% greater than the model-based estimates. The differences are even larger on regional scales. Such measurement-model differences are a combination of differences in aerosol amount (optical depth), single-scattering properties, surface albedo, and radiative transfer schemes (Yu et al., 2006). As discussed earlier, MODIS retrieved optical depths tend to be overestimated by about 10-15% due to the contamination of thin cirrus and clouds in

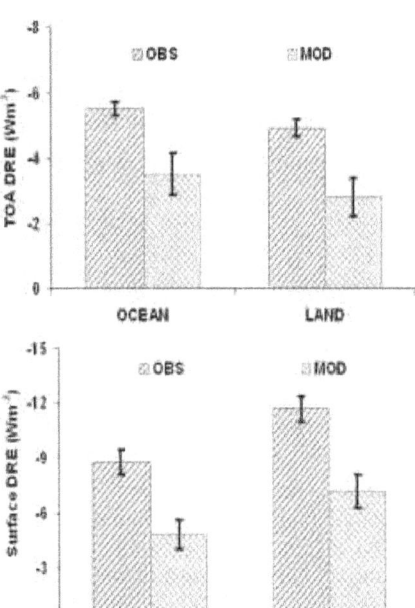

Figure 2.13. Summary of observation- and model-based (denoted as OBS and MOD, respectively) estimates of clear-sky, annual average DRF at the TOA and at the surface. The box and vertical bar represent median and standard error, respectively. (taken from Yu et al., 2006).

Table 2.5. Summary of seasonal and annual average clear-sky DRF (W m⁻²) at the TOA and the surface (SFC) over global **OCEAN** derived with different methods and data. Sources of data: **MODIS** (Remer & Kaufman, 2006), **MODIS_A** (Bellouin et al., 2005), **POLDER** (Boucher and Tanré, 2000; Bellouin et al., 2003), **CERES_A** and **CERES_B** (Loeb and Manalo-Smith, 2005), **CERES_C** (Zhang et al., 2005b), **MODIS_G, MISR_G, MO_GO, MO_MI_GO** (Yu et al., 2004; 2006), **SeaWiFS** (Chou et al., 2002), **GOCART** (Chin et al., 2002; Yu et al., 2004), **SPRINTARS** (Takemura et al., 2002), **GISS** (Koch and Hansen, 2005; Koch et al., 2006), **LMDZ-INCA** (Kinne et al., 2006; Schulz et al., 2006), **LMDZ-LOA** (Reddy et al., 2005a, b). Mean, median, standard deviation (σ), and standard error (ε) are calculated for observations (Obs) and model simulations (Mod) separately. The last row is the ratio of model median to observational median. (taken from Yu et al., 2006)

Products	DJF		MAM		JJA		SON		ANN	
	TOA	SFC	TOA	SFC	TOA	SFC	TOA	SFC	TOA	SFC
MODIS	-5.9		-5.8		-6.0		-5.8		-5.9	
MODIS_A*	-6.0	-8.2	-6.4	-8.9	-6.5	-9.3	-6.4	-8.9	-6.4	-8.9
CERES_A	-5.2		-6.1		-5.4		-5.1		-5.5	
CERES_B	-3.8		-4.3		-3.5		-3.6		-3.8	
CERES_C	-5.3		-5.4		-5.2				-5.3	
MODIS_G	-5.5	-9.1	-5.7	-10.4	-6.0	-10.6	-5.5	-9.8	-5.7	-10.0
MISR_G**	-6.4	-10.3	-6.5	-11.4	-7.0	-11.9	-6.3	-10.9	-6.5	-11.1
MO_GO	-4.9	-7.8	-5.1	-9.3	-5.4	-9.4	-5.0	-8.7	-5.1	-8.8
MO_MI_GO	-4.9	-7.9	-5.1	-9.2	-5.5	-9.5	-5.0	-8.6	-5.1	-8.7
POLDER	-5.7		-5.7		-5.8		-5.6		-5.7	
									-5.2***	-7.7***
SeaWiFS	-6.0	-6.6	-5.2	-5.8	-4.9	-5.6	-5.3	-5.7	-5.4	-5.9
Obs. Mean	-5.4	-8.3	-5.6	-9.2	-5.6	-9.4	-5.4	-8.8	-5.5	-8.7
Obs. Median	-5.5	-8.1	-5.7	-9.3	-5.5	-9.5	-5.4	-8.8	-5.5	-8.8
Obs. σ	0.72	1.26	0.64	1.89	0.91	2.10	0.79	1.74	0.70	1.65
Obs. ε	0.23	0.56	0.20	0.85	0.29	0.94	0.26	0.78	0.21	0.67
GOCART	-3.6	-5.7	-4.0	-7.2	-4.7	-8.0	-4.0	-6.8	-4.1	-6.9
SPRINTARS	-1.5	-2.5	-1.5	-2.5	-1.9	-3.3	-1.5	-2.5	-1.6	-2.7
GISS	-3.3	-4.1	-3.5	-4.6	-3.5	-4.9	-3.8	-5.4	-3.5	-4.8
LMDZ -INCA	-4.6	-5.6	-4.7	-5.9	-5.0	-6.3	-4.8	-5.5	-4.7	-5.8
LMDZ -LOA	-2.2	-4.1	-2.2	-3.7	-2.5	-4.4	-2.2	-4.1	-2.3	-4.1
Mod. Mean	-3.0	-4.4	-3.2	-4.8	-3.5	-5.4	-3.3	-4.9	-3.2	-4.9
Mod. Median	-3.3	-4.1	-3.5	-4.6	-3.5	-4.9	-3.8	-5.4	-3.5	-4.8
Mod. σ	1.21	1.32	1.31	1.84	1.35	1.82	1.36	1.63	1.28	1.6
Mod. ε	0.61	0.66	0.66	0.92	0.67	0.91	0.68	0.81	0.64	0.80
Mod./Obs.	0.60	0.51	0.61	0.50	0.64	0.52	0.70	0.61	0.64	0.55

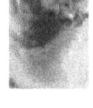

* High bias may result from adding the DRF of individual components to derive the total DRF (Bellouin et al., 2005).
** High bias most likely results from an overall overestimate of 20% in early post-launch MISR optical depth retrievals (Kahn et al., 2005).
*** Bellouin et al. (2003) use AERONET retrieval of aerosol absorption as a constraint to the method in Boucher and Tanré (2000), deriving aerosol direct radiative forcing both at the TOA and the surface.

general (Kaufman et al., 2005b). Such over-estimation of optical depth would result in a comparable overestimate of the aerosol direct radiative forcing. Other satellite AOD data may have similar contamination, which however has not yet been quantified. On the other hand, the observations may be measuring enhanced AOD and DRF due to processes not well represented in the models including humidification and enhancement of aerosols in the vicinity of clouds (Koren et al., 2007b).

From the perspective of model simulations, uncertainties associated with parameterizations of various aerosol processes and meteorological fields, as documented under the AEROCOM and Global Modeling Initiative (GMI) frameworks (Kinne et al., 2006; Textor et al., 2006; Liu et al., 2007), contribute to the large measurement-model and model-model discrepancies. Factors determining the AOD should be major reasons for the DRF discrepancy and the constraint of model AOD with well evaluated and bias reduced satellite AOD through a data assimilation approach can reduce the DRF discrepancy significantly. Other factors (such as model parameterization of surface reflectance, and model-satellite differences in single-scattering albedo and asymmetry factor due to satellite sampling bias toward cloud-free conditions) should also contribute, as evidenced by the existence of a large discrepancy in the radiative efficiency (Yu et al., 2006). Significant effort will be needed in the future to conduct comprehensive assessments.

On a global average, the measurement-based estimates of aerosol direct radiative forcing are 55-80% greater than the model-based estimates. The differences are even larger on regional scales.

Table 2.6. Summary of seasonal and annual average clear-sky DRF (W m⁻²) at the TOA and the surface (SFC) over global LAND derived with different methods and data. Sources of data: MODIS_G, MISR_G, MO_GO, MO_MI_GO (Yu et al., 2004, 2006), GOCART (Chin et al., 2002; Yu et al., 2004), SPRINTARS (Takemura et al., 2002), GISS (Koch and Hansen, 2005; Koch et al., 2006), LMDZ-INCA (Balkanski et al., 2007; Kinne et al., 2006; Schulz et al., 2006), LMDZ-LOA (Reddy et al., 2005a, b). Mean, median, standard deviation (σ), and standard error (ε) are calculated for observations (Obs) and model simulations (Mod) separately. The last row is the ratio of model median to observational median. (taken from Yu et al., 2006)

Products	DJF		MAM		JJA		SON		ANN	
	TOA	SFC	TOA	SFC	TOA	SFC	TOA	SFC	TOA	SFC
MODIS_G	-4.1	-9.1	-5.8	-14.9	-6.6	-17.4	-5.4	-12.8	-5.5	-13.5
MISR_G	-3.9	-8.7	-5.1	-13.0	-5.8	-14.6	-4.6	-10.7	-4.9	-11.8
MO_GO	-3.5	-7.5	-5.1	-12.9	-5.8	-14.9	-4.8	-10.9	-4.8	-11.6
MO_MI_GO	-3.4	-7.4	-4.7	-11.8	-5.3	-13.5	-4.3	-9.7	-4.4	-10.6
Obs. Mean	-3.7	-8.2	-5.2	-13.2	-5.9	-15.1	-4.8	-11.0	-4.9	-11.9
Obs. Median	-3.7	-8.1	-5.1	-13.0	-5.8	-14.8	-4.7	-10.8	-4.9	-11.7
Obs. σ	0.33	0.85	0.46	1.29	0.54	1.65	0.46	1.29	0.45	1.20
Obs. ε	0.17	0.49	0.26	0.74	0.31	0.85	0.27	0.75	0.26	0.70
GOCART	-2.9	-6.1	-4.4	-10.9	-4.8	-12.3	-4.3	-9.3	-4.1	-9.7
SPRINTARS	-1.4	-4.0	-1.5	-4.6	-2.0	-6.7	-1.7	-5.2	-1.7	-5.1
GISS	-1.6	-3.9	-3.2	-7.9	-3.6	-9.3	-2.5	-6.6	-2.8	-7.2
LMDZ-INCA	-3.0	-5.8	-4.0	-9.2	-6.0	-13.5	-4.3	-8.2	-4.3	-9.2
LMDZ-LOA	-1.3	-5.4	-1.8	-6.4	-2.7	-8.9	-2.1	-6.7	-2.0	-6.9
Mod. Mean	-2.0	-5.0	-3.0	-7.8	-3.8	-10.1	-3.0	-7.2	-3.0	-7.6
Mod. Median	-1.6	-5.4	-3.2	-7.9	-3.6	-9.3	-2.5	-6.7	-2.8	-7.2
Mod. σ	0.84	1.03	1.29	2.44	1.61	2.74	1.24	1.58	1.19	1.86
Mod. ε	0.42	0.51	0.65	1.22	0.80	1.37	0.62	0.79	0.59	0.93
Mod./Obs.	0.43	0.67	0.63	0.61	0.62	0.63	0.53	0.62	0.58	0.62

2.3.4. SATELLITE BASED ESTIMATES OF AN-THROPOGENIC COMPONENT OF AERO-SOL DIRECT RADIATIVE FORCING

Satellite instruments do not measure the aerosol chemical composition needed to discriminate anthropogenic from natural aerosol components. Because anthropogenic aerosols are predominantly sub-micron, the fine-mode fraction derived from POLDER, MODIS, or MISR might be used as a tool for deriving anthropogenic aerosol optical depth. This could provide a feasible way to conduct measurement-based estimates of anthropogenic component of aerosol direct radiative forcing (Kaufman et al., 2002a). Such method derives anthropogenic AOD from satellite measurements by empirically correcting contributions of natural sources (dust and maritime aerosol) to the sub-micron AOD (Kaufman et al., 2005a). The MODIS-based estimate of anthropogenic AOD is about 0.033 over oceans, consistent with model assessments of 0.030~0.036 even though the total AOD from MODIS is 25-40% higher than the models (Kaufman et al., 2005a). This accounts for $21\pm7\%$ of the MODIS-observed total aerosol optical depth, compared with about 33% of anthropogenic contributions estimated by the models. The anthropogenic fraction of AOD should be much larger over land (i.e., $47\pm9\%$ from a composite of several models) (Bellouin et al., 2005), comparable to the 40% estimated by Yu et al. (2006). Similarly, the non-spherical fraction from MISR or POLDER can be used to separate dust from spherical aerosol (Kahn et al., 2001; Kalashnikova and Kahn, 2006), providing another constraint for distinguishing anthropogenic from natural aerosols.

There have been several estimates of anthropogenic component of DRF in recent years. Table 2.7 lists such estimates of anthropogenic component of TOA DRF that are from model simulations (Schulz et al., 2006) and constrained to some degree by satellite observations (Kaufman et al., 2005a; Bellouin et al., 2005, 2008; Chung et al., 2005; Christopher et al., 2006; Matsui and Pielke, 2006; Yu et al., 2006; Quaas et al., 2008; Zhao et al., 2008b). The satellite-based clear-sky DRF by anthropogenic aerosols is estimated to be -1.1 ± 0.37 W m^{-2} over ocean, about a factor of 2 stronger than model simulated -0.6 W m^{-2}. Similar DRF estimates are rare over land, but a few studies do suggest that the anthropogenic

DRF over land is much more negative than that over ocean (Yu et al., 2006; Bellouin et al., 2005, 2008). On global average, the measurement-based estimate of anthropogenic DRF ranges from -0.9 to -1.9 W m^{-2}, again stronger than the model-based estimate of -0.8 W m^{-2}. Similar to DRF estimates for total aerosols, satellite-based estimates of anthropogenic component of DRF are rare over land.

On global average, anthropogenic aerosols are generally more absorptive than natural aerosols. As such the anthropogenic component of DRF is much more negative at the surface than at TOA. Several observation-constrained studies estimate that the global average, clear-sky, anthropogenic component of DRF at the surface ranges from -4.2 to -5.1 W m^{-2} (Yu et al., 2004; Bellouin et al., 2005; Chung et al., 2005; Matsui and Pielke, 2006), which is about a factor of 2 larger in magnitude than the model estimates (e.g., Reddy et al., 2005b).

Uncertainties in estimates of the anthropogenic component of aerosol DRF are greater than for the total aerosol, particularly over land. An uncertainty analysis (Yu et al., 2006) partitions the uncertainty for the global average anthropogenic DRF between land and ocean more or less evenly. Five parameters, namely fine-mode fraction (f_f) and anthropogenic fraction of fine-mode fraction (f_{af}) over both land and ocean, and τ over ocean, contribute nearly 80% of the overall uncertainty in the anthropogenic DRF estimate, with individual shares ranging from 13-20% (Yu et al., 2006). These uncertainties presumably represent a lower bound because the sources of error are assumed to be independent. Uncertainties associated with several parameters are also not well defined. Nevertheless, such uncertainty analysis is useful for guiding future research and documenting advances in understanding.

2.3.5. AEROSOL-CLOUD INTERACTIONS AND INDIRECT FORCING

Satellite views of the Earth show a planet whose albedo is dominated by dark oceans and vegetated surfaces, white clouds, and bright deserts. The bright white clouds overlying darker oceans or vegetated surface demonstrate the significant effect that clouds have on the Earth's radiative balance. Low clouds reflect

Uncertainties in estimates of the anthropogenic component of aerosol direct radiative forcing are greater than for the total aerosol, particularly over land.

Table 2.7. Estimates of anthropogenic components of aerosol optical depth (τ_{ant}) and clear-sky DRF at the TOA from model simulations (Schulz et al., 2006) and approaches constrained by satellite observations (Kaufman et al., 2005a; Bellouin et al., 2005, 2008; Chung et al., 2005; Yu et al., 2006; Christopher et al., 2006; Matsui and Pielke, 2006; Quaas et al., 2008; Zhao et al., 2008b).

Data Sources	Ocean		Land		Global		Estimated uncertainty or model diversity for DRF
	τ_{ant}	DRF (W m^{-2})	τ_{ant}	DRF (W m^{-2})	τ_{ant}	DRF (W m^{-2})	
Kaufman et al. (2005a)	0.033	-1.4					30%
Bellouin et al. (2005)	0.028	-0.8	0.13		0.062	-1.9	15%
Chung et al. (2005)						-1.1	
Yu et al. (2006)	0.031	-1.1	0.088	-1.8	0.048	-1.3	47% (ocean), 84% (land), and 62% (global)
Christopher et al. (2006)		-1.4					65%
Matsui and Pielke (2006)		-1.6					30°S-30°N oceans
Quaas et al. (2008)		-0.7		-1.8		-0.9	45%
Bellouin et al. (2008)	0.021	-0.6	0.107	-3.3	0.043	-1.3	Update to Bellouin et al. (2005) with MODIS Collection 5 data
Zhao et al. (2008b)		-1.25					35%
Schulz et al. (2006)	0.022	-0.59	0.065	-1.14	0.036	-0.77	30-40%; same emissions prescribed for all models

incoming sunlight back to space, acting to cool the planet, whereas high clouds can trap outgoing terrestrial radiation and act to warm the planet. In the Arctic, low clouds have also been shown to warm the surface (Garrett and Zhao, 2006). Changes in cloud cover, in cloud vertical development, and cloud optical properties will have strong radiative and therefore, climatic impacts. Furthermore, factors that change cloud development will also change precipitation processes. These changes may alter amounts, locations and intensities of local and regional rain and snowfall, creating droughts, floods and severe weather.

Cloud droplets form on a subset of aerosol particles called cloud condensation nuclei (CCN). In general, an increase in aerosol leads to an increase in CCN and an increase in drop concentration. Thus, for the same amount of liquid water in a cloud, more available CCN will result in a greater number but smaller size of

droplets (Twomey, 1977). A cloud with smaller but more numerous droplets will be brighter and reflect more sunlight to space, thus exerting a cooling effect. This is the first aerosol indirect radiative effect, or "albedo effect". The effectiveness of a particle as a CCN depends on its size and composition so that the degree to which clouds become brighter for a given aerosol perturbation, and therefore the extent of cooling, depends on the aerosol size distribution and its size-dependent composition. In addition, aerosol perturbations to cloud microphysics may involve feedbacks; for example, smaller drops are less likely to collide and coalesce; this will inhibit growth, suppressing precipitation, and possibly increasing cloud lifetime (Albrecht et al., 1989). In this case clouds may exert an even stronger cooling effect.

A distinctly different aerosol effect on clouds exists in thin Arctic clouds (LWP < 25 g m^{-2}) having low emissivity. Aerosol has been shown to increase the longwave emissivity in these clouds, thereby *warming* the surface (Lubin and Vogelman, 2006; Garrett and Zhao, 2006).

Some aerosol particles, particularly black carbon and dust, also act as ice nuclei (IN) and in so doing, modify the microphysical properties of mixed-phase and ice-clouds. An increase in IN will generate more ice crystals, which grow at the expense of water droplets due to the difference in vapor pressure over ice and water surfaces. The efficient growth of ice particles may increase the precipitation efficiency. In deep convective, polluted clouds there is a delay in the onset of freezing because droplets are smaller. These clouds may eventually precipitate, but only after higher altitudes are reached that result in taller cloud tops, more lightning and greater chance of severe weather (Rosenfeld and Lensky, 1998; Andreae et al., 2004). The present state of knowledge of the nature and abundance of IN, and ice formation in clouds is extremely poor. There is some observational evidence of aerosol influences on ice processes, but a clear link between aerosol, IN concentrations, ice crystal concentrations and growth to precipitation has not been established. This report therefore only peripherally addresses ice processes. More information can be found in a review by the WMO/IUGG International Aerosol-Precipitation Scientific Assessment (Levin and Cotton, 2008).

In addition to their roles as CCN and IN, aerosols also absorb and scatter light, and therefore they can change atmospheric conditions (temperature, stability, and surface fluxes) that influence cloud development and properties (Hansen et al., 1997; Ackerman et al., 2000). Thus, aerosols affect clouds through changing cloud droplet size distributions, cloud particle phase, and by changing the atmospheric environment of the cloud.

2.3.5A. REMOTE SENSING OF AEROSOL-CLOUD INTERACTIONS AND INDIRECT FORCING

The AVHRR satellite instruments have observed relationships between columnar aerosol loading, retrieved cloud microphysics, and cloud brightness over the Amazon Basin that are consistent with the theories explained above (Kaufman and Nakajima, 1993; Kaufman and Fraser, 1997; Feingold et al., 2001), but do not necessarily prove a causal relationship. Other studies have linked cloud and aerosol microphysical parameters or cloud albedo and droplet size using satellite data applied over the entire global oceans (Wetzel and Stowe, 1999; Nakajima et al., 2001; Han et al., 1998). Using these correlations with estimates of aerosol increase from the pre-industrial era, estimates of anthropogenic aerosol indirect radiative forcing fall into the range of -0.7 to -1.7 W m^{-2} (Nakajima et al., 2001).

Introduction of the more modern instruments (POLDER and MODIS) has allowed more detailed observations of relationships between aerosol and cloud parameters. Cloud cover can both decrease and increase with increasing aerosol loading (Koren et al., 2004; Kaufman et al., 2005c; Koren et al., 2005; Sekiguchi et al., 2003; Matheson et al., 2005; Yu et al., 2007). The same is true of LWP (Han et al., 2002; Matsui et al., 2006). Aerosol absorption appears to be an important factor in determining how cloud cover will respond to increased aerosol loading (Kaufman and Koren, 2006; Jiang and Feingold, 2006; Koren et al., 2008). Different responses of cloud cover to increased aerosol could also be correlated with atmospheric thermodynamic and moisture structure (Yu et al., 2007). Observations in the MODIS data show that aerosol loading correlates with enhanced convection and greater production of ice anvils in the summer Atlantic Ocean (Koren et al., 2005), which conflicts with previous results that used AVHRR and could not isolate convective systems from shallow clouds (Sekiguchi et al., 2003).

In recent years, surface-based remote sensing has also been applied to address aerosol effects on cloud microphysics. This method offers some interesting insights, and is complementary to the global satellite view. Surface remote sensing can only be applied at a limited number of locations, and therefore lacks the global satellite view. However, these surface stations yield high temporal resolution data and because they sample aerosol below, rather than adjacent to clouds they do not suffer from "cloud contamination". With the appropriate instrumentation (lidar) they can measure the local aerosol entering the clouds, rather than a column-integrated aerosol optical depth. Under well-mixed conditions, surface *in situ* aerosol measurements can be used. Surface remote-sensing studies are discussed in more

The present state of knowledge of the nature and abundance of ice nuclei and ice formation in clouds is extremely poor.

detail below, although the main science issues are common to satellite remote sensing.

Feingold et al. (2003) used data collected at the ARM Southern Great Plains (SGP) site to allow simultaneous retrieval of aerosol and cloud properties. A combination of a Doppler cloud radar and a microwave radiometer was used to retrieve cloud drop effective radius r_e profiles in non-precipitating (radar reflectivity $Z < -17$ dBZ), ice-free clouds. Simultaneously, sub-cloud aerosol extinction profiles were measured with a lidar to quantify the response of drop sizes to changes in aerosol properties. Cloud data were binned according to liquid water path (LWP) as measured with a micro-wave radiometer, consistent with Twomey's (1977) conceptual view of the aerosol impact on cloud microphysics. With high temporal/spatial resolution data (on the order of 20's or 100's of meters), realizations of aerosol-cloud interactions at the large eddy scale were obtained, and quantified in terms of the relative decrease in r_e in response to a relative increase in aerosol extinction ($d\ln r_e/d\ln$ extinction), as shown in Figure 2.14. Examining the dependence in this way reduces reliance on absolute measures of cloud and aerosol parameters and minimizes sensitivity to measurement error, provided errors are unbiased. This formulation permitted these responses to be related to cloud micro-physical theory. Restricting the examination to updrafts only (as determined from the radar Doppler signal) permitted examination of the role of updraft in determining the response of r_e to changes in aerosol (via changes in drop number concentration N_d). Analysis of data from 7 days showed that turbulence intensifies the aerosol impact on cloud microphysics.

In addition to radar/microwave radiometer retrievals of aerosol and cloud properties, measurements of cloud optical depth by sur-face based radiometers such as the MFRSR (Michalsky et al., 2001) have been used in combination with measurements of cloud LWP by microwave radiometer to measure an aver-age value of r_e during daylight when the solar elevation angle is sufficiently high (Min and Harrison, 1996). Using this retrieval, Kim et al. (2003) performed analyses of the r_e response to changes in aerosol at the same continen-tal site, using a surface measurement of the aerosol light scattering coefficient instead of

using extinction near cloud base as a proxy for CCN. Variance in LWP was shown to explain most of the variance in cloud optical depth, exacerbating detection of an aerosol effect. Although a decrease in r_e was observed with increasing scattering coefficient, the relation was not strong, indicative of other influences on r_e and/or decoupling between the surface and cloud layer. A similar study was conducted by Garrett et al. (2004) at a location in the Arctic.

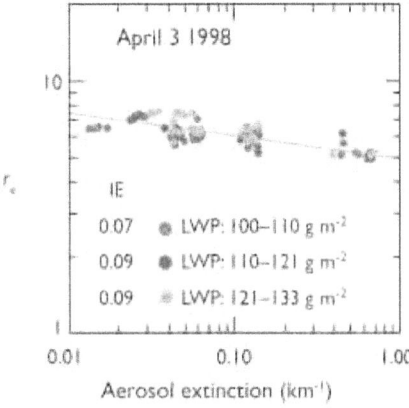

Figure 2.14. Scatter plots showing mean cloud drop effective radius (r_e) vs. aerosol extinction coef-ficient (unit: km^{-1}) for various liquid water path (LWP) bands on April 3, 1998 at ARM SGP site (adapted from Feingold et al., 2003).

They suggested that summertime Arctic clouds are more sensitive to aerosol perturbations than clouds at lower latitudes. The advantage of the MFRSR/microwave radiometer combination is that it derives r_e from cloud optical depth and LWP and it is not as sensitive to large drops as the radar is. A limitation is that it can be applied only to clouds with extensive horizontal cover during daylight hours.

More recent data analyses by Feingold et al. (2006), Kim et al. (2008) and McComiskey et al. (2008b) at a variety of locations, and modeling work (Feingold, 2003) have investigated (i) the use of different proxies for cloud condensation nuclei, such as the light scattering coefficient and aerosol index; (ii) sensitivity of cloud microphysical/optical properties to control-ling factors such as aerosol size distribution, entrainment, LWP, and updraft velocity; (iii) the effect of optical- as opposed to radar-retrievals of drop size; and (iv) spatial heterogeneity. These studies have reinforced the importance of LWP and vertical velocity as controlling parameters. They have also begun to reconcile the reasons for the large discrepancies between various approaches, and platforms (satellite, air-craft *in situ*, and surface-based remote sensing). These investigations are important because satellite measurements that use a similar approach are being employed in GCMs to represent the albedo indirect effect (Quaas and Boucher, 2005). In fact, the weakest albedo indirect effect in IPCC (2007) derives from satellite measurements that have very weak responses of r_e to changes in aerosol. The relationship between these aerosol-cloud microphysical responses and cloud radiative forcing has been examined by McComiskey and Feingold (2008). They showed that for plane-parallel clouds, a typical uncertainty in the logarithmic gradient of a r_e-aerosol relationship of 0.05 results in a local forcing error of -3 to -10 W m^{-2}, depending on the aerosol perturbation. This sensitivity reinforces the importance of adequate quantification of aerosol effects on cloud microphysics to assessment of the radiative forcing, i.e., the indirect effect. Quantification of these effects from remote sensors is exacerbated by measurement errors. For example, LWP is measured to an accuracy of 25 g m^{-2} at best, and since it is the thinnest clouds (i.e., low LWP) that are most susceptible (from a radiative forcing perspective) to changes in aerosol, this measurement uncertainty represents a significant uncertainty in whether the observed response is related to aerosol, or to differences in LWP. The accuracy and spatial resolution of satellite-based LWP measurements is much poorer and this represents a significant challenge. In some cases important measurements are simply absent, e.g., updraft is not measured from satellite-based remote sensors.

Finally, cloud radar data from CloudSat, along with the A-train aerosol data, is providing great opportunity for inferring aerosol effects on precipitation (e.g., Stephens and Haynes, 2007). The aerosol effect on precipitation is far more complex than the albedo effect because the instantaneous view provided by satellites makes it difficult to establish causal relationships.

2.3.5B. IN SITU STUDIES OF AEROSOL-CLOUD INTERACTIONS

In situ observations of aerosol effects on cloud microphysics date back to the 1950s and 1960s (Gunn and Phillips, 1957; Squires, 1958; Warner, 1968; Warner and Twomey, 1967; Radke et al., 1989; Leaitch et al., 1992; Brenguier et al., 2000; to name a few). These studies showed that high concentrations of CCN from anthropogenic sources, such as industrial pollution

or the burning of sugarcane, can increase cloud droplet number concentration N_d, thus increasing cloud microphysical stability and potentially reducing precipitation efficiency. As in the case of remote sensing studies, the causal link between aerosol perturbations and cloud microphysical responses (e.g., r_e or N_d) is much better established than the relationship between aerosol and changes in cloud fraction, LWC, and precipitation (see also Levin and Cotton, 2008).

In situ cloud measurements are usually regarded as "ground truth" for satellite retrievals but in fact there is considerable uncertainty in measured parameters such liquid water content (LWC), and size distribution, which forms the basis of other calculations such as drop concentration, r_e and extinction. It is not uncommon to see discrepancies in LWC on the order of 50% between different instruments, and cloud drop size distributions are difficult to measure, particularly for droplets < 10 μm where Mie scattering oscillations generate ambiguities in drop size. Measurement uncertainty in r_e from *in situ* probes is assessed, for horizontally homogeneous clouds, to be on the order of 15-20%, compared to 10% for MODIS and 15-20% for other spectral measurements (Feingold et al., 2006). As with remote measurements it is prudent to consider relative (as opposed to absolute) changes in cloud microphysics related to relative changes in aerosol. An added consideration is that *in situ* measurements typically represent a very small sample of the atmosphere akin to a thin pencil line through a large volume. For an aircraft flying at 100 m s^{-1} and sampling at 1 Hz, the sample volume is on the order of 10 cm^3. The larger spatial sampling of remote sensing has the advantage of being more representative but it removes small-scale (i.e., sub sampling-volume) variability, and therefore may obscure important cloud processes.

Measurements at a wide variety of locations around the world have shown that increases in aerosol concentration lead to increases in N_d. However the rate of this increase is highly variable and always sub-linear, as exemplified by the compilation of data in Ramanathan et al. (2001a). This is because, as discussed previously, N_d is a function of numerous parameters in addition to aerosol number concentration, including size distribution, updraft veloc-

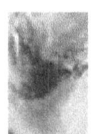

The aerosol effect on precipitation is far more complex than the albedo effect because the instantaneous view provided by satellites makes it difficult to establish causal relationships.

ity (Leaitch et al., 1996), and composition. In stratocumulus clouds, characterized by relatively low vertical velocity (and low supersaturation) only a small fraction of particles can be activated whereas in vigorous cumulus clouds that have high updraft velocities, a much larger fraction of aerosol particles is activated. Thus the ratio of N_d to aerosol particle number concentration is highly variable.

In recent years there has been a concerted effort to reconcile measured N_d concentrations with those calculated based on observed aerosol size and composition, as well as updraft velocity. These so-called "closure experiments" have demonstrated that on average, agreement in N_d between these approaches is on the order of 20% (e.g., Conant et al., 2004). This provides confidence in theoretical understanding of droplet activation, however, measurement accuracy is not high enough to constrain the aerosol composition effects that have magnitudes < 20%.

One exception to the rule that more aerosol particles result in larger N_d is the case of giant CCN (sizes on the order of a few microns), which, in concentrations on the order of 1 cm^{-3} (i.e., ~ 1% of the total concentration) can lead to significant suppression in cloud supersaturation and reductions in N_d (O'Dowd et al., 1999). The measurement of these large particles is difficult and hence the importance of this effect is hard to assess. These same giant CCN, at concentrations as low as 1/liter, can significantly affect the initiation of precipitation in moderately polluted clouds (Johnson, 1982) and in so doing alter cloud albedo (Feingold et al., 1999).

The most direct link between the remote sensing of aerosol-cloud interactions discussed in section 2.3.5.1 and *in situ* observations is via observations of relationships between drop concentration N_d and CCN concentration. Theory shows that if r_e-CCN relationships are calculated at constant LWP or LWC, their logarithmic slope is -1/3 that of the N_d-CCN logarithmic slope (i.e., $d\ln r_e/d\ln CCN = -1/3\ d\ln N_d/d\ln CCN$). In general, N_d-CCN slopes measured *in situ* tend to be stronger than equivalent slopes obtained from remote sensing – particularly in the case of satellite remote sensing (McComiskey and Feingold 2008). There are a number of reasons for this: (i) *in situ* measurements focus on smaller spatial scales and are more likely to observe the droplet activation process as opposed to remote sensing that incorporates larger spatial scales and includes other processes such as drop coalescence that reduce N_d, and therefore the slope of the N_d-CCN relationship (McComiskey et al., 2008b). (ii) Satellite remote sensing studies typically do not sort their data by LWP, and this has been shown to reduce the magnitude of the r_e-CCN response (Feingold, 2003).

In conclusion, observational estimates of aerosol indirect radiative forcings are still in their infancy. Effects on cloud microphysics that result in cloud brightening have to be considered along with effects on cloud lifetime, cover, vertical development and ice production. For *in situ* measurements, aerosol effects on cloud microphysics are reasonably consistent (within ~ 20%) with theory but measurement uncertainties in remote sensing of aerosol effects on clouds, as well as complexity associated with three-dimensional radiative transfer, result in considerable uncertainty in radiative forcing. The higher order indirect effects are poorly understood and even the sign of the microphysical response and forcing may not always be the same. Aerosol type and specifically the absorption properties of the aerosol may cause different cloud responses. Early estimates of observationally based aerosol indirect forcing range from -0.7 to -1.7 W m^{-2} (Nakajima et al., 2001) and -0.6 to -1.2 W m^{-2} (Sekiguchi et al., 2003), depending on the estimate for aerosol increase from pre-industrial times and whether aerosol effects on cloud fraction are also included in the estimate.

2.4. Outstanding Issues
Despite substantial progress, as summarized in section 2.2 and 2.3, most measurement-based studies so far have concentrated on influences produced by the sum of natural and anthropogenic aerosols on solar radiation under clear sky conditions. Important issues remain:
- Because accurate measurements of aerosol absorption are lacking and land surface reflection values are uncertain, DRF estimates over land and at the ocean surface are less well constrained than the estimate of TOA DRF over ocean.
- Current estimates of the anthropogenic component of aerosol direct radiative forcing have large uncertainties, especially over land.

For *in situ* measurements, aerosol effects on cloud microphysics are reasonably consistent with theory but measurement uncertainties in remote sensing of aerosol effects on clouds result in considerable uncertainty in radiative forcing.

- Because there are very few measurements of aerosol absorption vertical distribution, mainly from aircraft during field campaigns, estimates of direct radiative forcing of above-cloud aerosols and profiles of atmospheric radiative heating induced by aerosol absorption are poorly constrained.
- There is a need to quantify aerosol impacts on thermal infrared radiation, especially for dust.
- The diurnal cycle of aerosol direct radiative forcing cannot be adequately characterized with currently available, sun-synchronous, polar orbiting satellite measurements.
- Measuring aerosol, cloud, and ambient meteorology contributions to indirect radiative forcing remains a major challenge.
- Long-term aerosol trends and their relationship to observed surface solar radiation changes are not well understood.

The current status and prospects for these areas are briefly discussed below.

Measuring aerosol absorption and single-scattering albedo: Currently, the accuracy of both *in situ* and remote sensing aerosol SSA measurements is generally ±0.03 at best, which implies that the inferred accuracy of clear sky aerosol DRF would be larger than 1 W m^{-2} (see Chapter 1). Recently developed photoacoustic (Arnott et al., 1997) and cavity ring down extinction cell (Strawa et al., 2002) techniques for measuring aerosol absorption produce SSA with improved accuracy over previous methods. However, these methods are still experimental, and must be deployed on aircraft. Aerosol absorption retrievals from satellites using the UV-technique have large uncertainties associated with its sensitivity to the height of the aerosol layer(s) (Torres et al., 2005), and it is unclear how the UV results can be extended to visible wavelengths. Views in and out of sunglint can be used to retrieve total aerosol extinction and scattering, respectively, thus constraining aerosol absorption over oceans (Kaufman et al., 2002b). However, this technique requires retrievals of aerosol scattering properties, including the real part of the refractive index, well beyond what has so far been demonstrated from space. In summary, there is a need to pursue a better understanding of the uncertainty in SSA from both *in situ* measurements and remote sensing retrievals and,

with this knowledge, to synthesize different data sets to yield a characterization of aerosol absorption with well-defined uncertainty (Leahy et al., 2007). Laboratory studies of aerosol absorption of specific known composition are also needed to interpret *in situ* measurements and remote sensing retrievals and to provide updated database of particle absorbing properties for models.

Estimating the aerosol direct radiative forcing over land: Land surface reflection is large, heterogeneous, and anisotropic, which complicates aerosol retrievals and DRF determination from satellites. Currently, the aerosol retrievals over land have relatively lower accuracy than those over ocean (Section 2.2.5) and satellite data are rarely used alone for estimating DRF over land (Section 2.3). Several issues need to be addressed, such as developing appropriate angular models for aerosols over land (Patadia et al., 2008) and improving land surface reflectance characterization. MODIS and MISR measure land surface reflection wavelength dependence and angular distribution at high resolution (Moody et al., 2005; Martonchik et al., 1998b; 2002). This offers a promising opportunity for inferring the aerosol direct radiative forcing over land from satellite measurements of radiative fluxes (e.g., CERES) and from critical reflectance techniques (Fraser and Kaufman, 1985; Kaufman, 1987). The aerosol direct radiative forcing over land depends strongly on aerosol absorption and improved measurements of aerosol absorption are required.

Distinguishing anthropogenic from natural aerosols: Current estimates of anthropogenic components of AOD and direct radiative forcing have larger uncertainties than total aerosol optical depth and direct radiative forcing, particularly over land (see Section 2.3.4), because of relatively large uncertainties in the retrieved aerosol microphysical properties (see Section 2.2). Future measurements should focus on improved retrievals of such aerosol properties as size distribution, particle shape, and absorption, along with algorithm refinement for better aerosol optical depth retrievals. Coordinated *in situ* measurements offer a promising avenue for validating and refining satellite identification of anthropogenic aerosols (Anderson et al., 2005a, 2005b). For satellite-based aerosol type characterization, it is sometimes assumed that

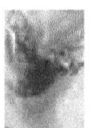

There is a need to pursue a better understanding of the uncertainty in single scattering albedo from both *in situ* measurements and remote sensing retrievals.

all biomass-burning aerosol is anthropogenic and all dust aerosol is natural (Kaufman et al., 2005a). The better determination of anthropogenic aerosols requires a quantification of biomass burning ignited by lightning (natural origin) and mineral dust due to human induced changes of land cover/land use and climate (anthropogenic origin). Improved emissions inventories and better integration of satellite observations with models seem likely to reduce the uncertainties in aerosol source attribution.

Profiling the vertical distributions of aerosols: Current aerosol profile data are far from adequate for quantifying the aerosol radiative forcing and atmospheric response to the forcing. The data have limited spatial and temporal coverage, even for current spaceborne lidar measurements. Retrieving aerosol extinction profile from lidar measured attenuated backscatter is subject to large uncertainties resulting from aerosol type characterization. Current space-borne Lidar measurements are also not sensitive to aerosol absorption. Because of lack of aerosol vertical distribution observations, the estimates of DRF in cloudy conditions and dust DRF in the thermal infrared remain highly uncertain (Schulz et al., 2006; Sokolik et al., 2001; Lubin et al., 2002). It also remains challenging to constrain the aerosol-induced atmospheric heating rate increment that is essential for assessing atmospheric responses to the aerosol radiative forcing (e.g., Yu et al., 2002; Feingold et al., 2005; Lau et al., 2006). Progress in the foreseeable future is likely to come from (1) better use of existing, global, space-based backscatter lidar data to constrain model simulations, and (2) deployment of new instruments, such as high-spectral-resolution lidar (HSRL), capable of retrieving both extinction and backscatter from space. The HSRL lidar system will be deployed on the EarthCARE satellite mission tentatively scheduled for 2013 (http://asimov/esrin.esi.it/esaLP/ASESMYN-W9SC_Lpearthcare_1.html).

Characterizing the diurnal cycle of aerosol direct radiative forcing: The diurnal variability of aerosol can be large, depending on location and aerosol type (Smirnov et al., 2002), especially in wildfire situations, and in places where boundary layer aerosols hydrate or otherwise change significantly during the day. This cannot be captured by currently available, sun-synchronous, polar orbiting satellites. Geostationary satellites provide adequate time resolution (Christopher and Zhang, 2002; Wang et al., 2003), but lack the information required to characterize aerosol types. Aerosol type information from low earth orbit satellites can help improve accuracy of geostationary satellite aerosol retrievals (Costa et al., 2004a, 2004b). For estimating the diurnal cycle of aerosol DRF, additional efforts are needed to adequately characterize the anisotropy of surface reflection (Yu et al., 2004) and daytime variation of clouds.

Studying aerosol-cloud interactions and indirect radiative forcing: Remote sensing estimates of aerosol indirect forcing are still rare and uncertain. Improvements are needed for both aerosol characterization and measurements of cloud properties, precipitation, water vapor, and temperature profiles. Basic processes still need to be understood on regional and global scales. Remote sensing observations of aerosol-cloud interactions and aerosol indirect forcing are for the most part based on simple correlations among variables, from which cause-and-effects cannot be deduced. One difficulty in inferring aerosol effects on clouds from the observed relationships is separating aerosol from meteorological effects, as aerosol loading itself is often correlated with the meteorology. In addition, there are systematic errors and biases in satellite aerosol retrievals for partly cloud-filled scenes. Stratifying aerosol and cloud data by liquid water content, a key step in quantifying the albedo (or first) indirect effect, is usually missing. Future work will need to combine satellite observations with *in situ* validation and modeling interpretation. A methodology for integrating observations (*in situ* and remote) and models at the range of relevant temporal/spatial scales is crucial to improve understanding of aerosol indirect effects and aerosol-cloud interactions.

Quantifying long-term trends of aerosols at regional scales: Because secular changes are subtle, and are superposed on seasonal and other natural variability, this requires the construction of consistent, multi-decadal records of climate-quality data. To be meaningful, aerosol trend analysis must be performed on a regional basis. Long-term trends of aerosol optical depth have been studied using measurements from surface remote sensing stations (e.g., Hoyt and

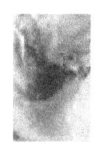

Remote sensing estimates of aerosol indirect forcing are still rare and uncertain. Improvements are needed for both aerosol characterization and measurements of cloud properties, precipitation, water vapor, and temperature profiles.

To be meaningful, aerosol trend analysis must be performed on a regional basis.

Frohlich, 1983; Augustine et al., 2008; Luo et al., 2001) and historic satellite sensors (Massie et al., 2004; Mishchenko et al., 2007a; Mishchenko and Geogdzhayev, 2007; Zhao et al., 2008a). An emerging multi-year climatology of high quality AOD data from modern satellite sensors (e.g., Remer et al., 2008; Kahn et al., 2005a) has been used to examine the interannual variations of aerosol (e.g., Koren et al., 2007a, Mishchenko and Geogdzhayev, 2007) and contribute significantly to the study of aerosol trends. Current observational capability needs to be continued to avoid any data gaps. A synergy of aerosol products from historical, modern and future sensors is needed to construct as long a record as possible. Such a data synergy can build upon understanding and reconciliation of AOD differences among different sensors or platforms (Jeong et al., 2005). This requires overlapping data records for multiple sensors. A close examination of relevant issues associated with individual sensors is urgently needed, including sensor calibration, algorithm assumptions, cloud screening, data sampling and aggregation, among others.

Linking aerosol long-term trends with changes of surface solar radiation: Analysis of the long-term surface solar radiation record suggests significant trends during past decades (e.g., Stanhill and Cohen, 2001; Wild et al., 2005; Pinker et al., 2005; Alpert et al., 2005). Although a significant and widespread decline in surface total solar radiation (the sum of direct and diffuse irradiance) occurred up to 1990 (so-called solar dimming), a sustained increase has been observed during the subsequent decade. Speculation suggests that such trends result from decadal changes of aerosols and the interplay of aerosol direct and indirect radiative forcing (Stanhill and Cohen, 2001; Wild et al., 2005; Streets et al., 2006a; Norris and Wild, 2007; Ruckstuhl et al., 2008). However, reliable observations of aerosol trends are required test these ideas. In addition to aerosol optical depth, changes in aerosol composition must also be quantified, to account for changing industrial practices, environmental regulations, and biomass burning emissions (Novakov et al., 2003; Streets et al., 2004; Streets and Aunan et al., 2005). Such compositional changes will affect the aerosol SSA and size distribution, which in turn will affect the surface solar radiation (e.g., Qian et

al., 2007). However such data are currently rare and subject to large uncertainties. Finally, a better understanding of aerosol-radiation-cloud interactions and trends in cloudiness, cloud albedo, and surface albedo is badly needed to attribute the observed radiation changes to aerosol changes with less ambiguity.

2.5. Concluding Remarks

Since the concept of aerosol-radiation-climate interactions was first proposed around 1970, substantial progress has been made in determining the mechanisms and magnitudes of these interactions, particularly in the last ten years. Such progress has greatly benefited from significant improvements in aerosol measurements and increasing sophistication of model simulations. As a result, knowledge of aerosol properties and their interaction with solar radiation on regional and global scales is much improved. Such progress plays a unique role in the definitive assessment of the global anthropogenic radiative forcing, as *"virtually certainly positive"* in IPCC AR4 (Haywood and Schulz, 2007).

In situ **measurements of aerosols:** New *in situ* instruments such as aerosol mass spectrometers, photoacoustic techniques, and cavity ring down cells provide high accuracy and fast time resolution measurements of aerosol chemical and optical properties. Numerous focused field campaigns and the emerging ground-based aerosol networks are improving regional aerosol chemical, microphysical, and radiative property characterization. Aerosol closure studies of different measurements indicate that measurements of submicrometer, spherical sulfate and carbonaceous particles have a much better accuracy than that for dust-dominated aerosol. The accumulated comprehensive data sets of regional aerosol properties provide a rigorous "test bed" and strong constraint for satellite retrievals and model simulations of aerosols and their direct radiative forcing.

Remote sensing measurements of aerosols: Surface networks, covering various aerosol regimes around the globe, have been measuring aerosol optical depth with an accuracy of 0.01~0.02, which is adequate for achieving the accuracy of 1 W m^{-2} for cloud-free TOA DRF. On the other hand, aerosol microphysical properties retrieved from these networks, especially

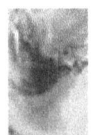

A better understanding of aerosol-radiation-cloud interactions and trends in cloudiness, cloud albedo, and surface albedo is badly needed to attribute the observed radiation changes to aerosol changes with less ambiguity.

SSA, have relatively large uncertainties and are only available in very limited conditions. Current satellite sensors can measure AOD with an accuracy of about 0.05 or 15 to 20% in most cases. The implementation of multi-wavelength, multi-angle, and polarization measuring capabilities has also made it possible to measure particle properties (size, shape, and absorption) that are essential for characterizing aerosol type and estimating anthropogenic component of aerosols. However, these microphysical measurements are more uncertain than AOD measurements.

Observational estimates of clear-sky aerosol direct radiative forcing: Closure studies based on focused field experiments reveal DRF uncertainties of about 25% for sulfate/carbonaceous aerosol and 60% for dust at regional scales. The high-accuracy of MODIS, MISR and POLDER aerosol products and broadband flux measurements from CERES make it feasible to obtain observational constraints for aerosol TOA DRF at a global scale, with relaxed requirements for measuring particle microphysical properties. Major conclusions from the assessment are:

- A number of satellite-based approaches consistently estimate the clear-sky diurnally averaged TOA DRF (on solar radiation) to be about -5.5±0.2 W m^{-2} (mean ± standard error from various methods) over global ocean. At the ocean surface, the diurnally averaged DRF is estimated to be -8.7±0.7 W m^{-2}. These values are calculated for the difference between today's measured total aerosol (natural plus anthropogenic) and the absence of all aerosol.
- Overall, in comparison to that over ocean, the DRF estimates over land are more poorly constrained by observations and have larger uncertainties. A few satellite retrieval and satellite-model integration yield the over-land clear-sky diurnally averaged DRF of -4.9±0.7 W m^{-2} and -11.8±1.9 W m^{-2} at the TOA and surface, respectively. These values over land are calculated for the difference between total aerosol and the complete absence of all aerosol.
- Use of satellite measurements of aerosol microphysical properties yields that on a global ocean average, about 20% of AOD is contributed by human activities and the clear-sky TOA DRF by anthropogenic aerosols is -1.1±0.4 W m^{-2}. Similar DRF estimates are

rare over land, but a few measurement-model integrated studies do suggest much more negative DRF over land than over ocean.
- These satellite-based DRF estimates are much greater than the model-based estimates, with differences much larger at regional scales than at a global scale.

Measurements of aerosol-cloud interactions and indirect radiative forcing: *In situ* measurement of cloud properties and aerosol effects on cloud microphysics suggest that theoretical understanding of the activation process for water cloud is reasonably well-understood. Remote sensing of aerosol effects on droplet size associated with the albedo effect tends to underestimate the magnitude of the response compared to *in situ* measurements. Recent efforts trace this to a combination of lack of stratification of data by cloud water, the relatively large spatial scale over which measurements are averaged (which includes variability in cloud fields, and processes that obscure the aerosol-cloud processes), as well as measurement uncertainties (particularly in broken cloud fields). It remains a major challenge to infer aerosol number concentrations from satellite measurements. The present state of knowledge of the nature and abundance of IN, and ice formation in clouds is extremely poor.

Despite the substantial progress in recent decades, several important issues remain, such as measurements of aerosol size distribution, particle shape, absorption, and vertical profiles, and the detection of aerosol long-term trend and establishment of its connection with the observed trends of solar radiation reaching the surface, as discussed in section 2.4. Furthering the understanding of aerosol impacts on climate requires a coordinated research strategy to improve the measurement accuracy and use the measurements to validate and effectively constrain model simulations. Concepts of future research in measurements are discussed in Chapter 4 "Way Forward".

The high-accuracy of satellite measurements makes it feasible to obtain observational constraints for aerosol top-of-atmosphere direct radiative forcing at a global scale.

Furthering the understanding of aerosol impacts on climate requires a coordinated research strategy to improve the measurement accuracy and to constrain/validate models with measurements.

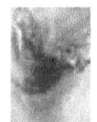

Sampling the Arcic Haze. Pollution and smoke aerosols can travel long distances, from mid-latitudes to the Arctic, causing "Arctic Haze". Photo taken from the NASA DC-8 aircraft during the ARCTAS field experiment over Alaska in April 2008. Credit: Mian Chin, NASA.

Modeling the Effects of Aerosols on Climate

Lead Authors: David Rind, NASA GISS; Mian Chin, NASA GSFC; Graham Feingold, NOAA ESRL; David G. Streets, DOE ANL
Contributing Authors: Ralph A. Kahn, NASA GSFC; Stephen E. Schwartz, DOE BNL; Hongbin Yu, NASA GSFC/UMBC

3.1. Introduction

The IPCC Fourth Assessment Report (AR4) (IPCC, 2007) concludes that man's influence on the warming climate is in the category of "very likely". This conclusion is based on, among other things, the ability of models to simulate the global and, to some extent, regional variations of temperature over the past 50 to 100 years. When anthropogenic effects are included, the simulations can reproduce the observed warming (primarily for the past 50 years); when they are not, the models do not get very much warming at all. In fact, all of the models runs for the IPCC AR4 assessment (more than 20) produce this distinctive result, driven by the greenhouse gas increases that have been observed to occur.

These results were produced in models whose average global warming associated with a doubled CO_2 forcing of 4 W m^{-2} was about 3°C. This translates into a climate sensitivity (surface temperature change per forcing) of about 0.75°C/(W m^{-2}). The determination of climate sensitivity is crucial to projecting the future impact of increased greenhouse gases, and the credibility of this projected value relies on the ability of these models to simulate the observed temperature changes over the past century. However, in producing the observed temperature trend in the past, the models made use of very uncertain aerosol forcing. The greenhouse gas change by itself produces warming in models that exceeds that observed by some 40% on average (IPCC, 2007). Cooling associ-

ated with aerosols reduces this warming to the observed level. Different climate models use differing aerosol forcings, both direct (aerosol scattering and absorption of short and long-wave radiation) and indirect (aerosol effect on cloud cover reflectivity and lifetime), whose magnitudes vary markedly from one model to the next. Kiehl (2007) using nine of the IPCC (2007) AR4 climate models found that they had a factor of three forcing differences in the aerosol contribution for the 20th century. The differing aerosol forcing is the prime reason why models whose climate sensitivity varies by almost a factor of three can produce the observed trend. It was thus concluded that the uncertainty in IPCC (2007) anthropogenic climate simulations for the past century should really be much greater than stated (Schwartz et al., 2007; Kerr, 2007), since, in general, models with low/high sensitivity to greenhouse warming used weaker/stronger aerosol cooling to obtain the same temperature response (Kiehl, 2007). Had the situation been reversed and the low/high sensitivity models used strong/weak aerosol forcing, there would have been a greater divergence in model simulations of the past century.

Therefore, the fact that a model has accurately reproduced the global temperature change in the past does not imply that its future forecast is accurate. This state of affairs will remain until a firmer estimate of radiative forcing (RF) by aerosols, in addition to that by greenhouse gases, is available.

The uncertainty in anthropogenic climate simulation for the past century should be much greater than stated, since models with low/high sensitivity to greenhouse warming used weaker/stronger aerosol cooling to obtain the same temperature response.

Two different approaches are used to assess the aerosol effect on climate. "Forward modeling" studies incorporate different aerosol types and attempt to explicitly calculate the aerosol RF. From this approach, IPCC (2007) concluded that the best estimate of the global aerosol direct RF (compared with preindustrial times) is -0.5 (-0.9 to -0.1) W m-2 (see Figure 1.3, Chapter 1). The RF due to the cloud albedo or brightness effect (also referred to as first indirect or Twomey effect) is estimated to be -0.7 (-1.8 to -0.3) W m-2. No estimate was specified for the effect associated with cloud lifetime. The total negative RF due to aerosols according to IPCC (2007) estimates (see Figure 1.3 in Chapter 1) is then -1.3 (-2.2 to -0.5) W m-2. In comparison, the positive radiative forcing (RF) from greenhouse gases (including tropospheric ozone) is estimated to be +2.9 ± 0.3 W m-2; hence tropospheric aerosols reduce the influence from greenhouse gases by about 45% (15-85%). This approach however inherits large uncertainties in aerosol amount, composition, and physical and optical properties in modeling of atmospheric aerosols. The consequences of these uncertainties are discussed in the next section.

The other method of calculating aerosol forcing is called the "inverse approach" – it is assumed that the observed climate change is primarily the result of the known climate forcing contributions. If one further assumes a particular climate sensitivity (or a range of sensitivities), one can determine what the total forcing had to be to produce the observed temperature change. The aerosol forcing is then deduced as a residual after subtraction of the greenhouse gas forcing along with other known forcings from the total value. Studies of this nature come up with aerosol forcing ranges of -0.6 to -1.7 W m-2 (Knutti et al., 2002, 2003; IPCC AR4 Chap.9); -0.4 to -1.6 W m-2 (Gregory et al., 2002); and -0.4 to -1.4 W m-2 (Stott et al., 2006). This approach however provides a bracket of the possible range of aerosol forcing without the assessment of current knowledge of the complexity of atmospheric aerosols.

This chapter reviews the current state of aerosol RF in the global models and assesses the uncertainties in these calculations. First representation of aerosols in the forward

global chemistry and transport models and the diversity of the model simulated aerosol fields are discussed; then calculation of the aerosol direct and indirect effects in the climate models is reviewed; finally the impacts of aerosols on climate model simulations and their implications are assessed.

3.2. Modeling of Atmospheric Aerosols

The global aerosol modeling capability has developed rapidly in the past decade. In the late 1990s, there were only a few global models that were able to simulate one or two aerosol components, but now there are a few dozen global models that simulate a comprehensive suite of aerosols in the atmosphere. As introduced in Chapter 1, aerosols consist of a variety of species including dust, sea salt, sulfate, nitrate, and carbonaceous aerosols (black and organic carbon) produced from natural and man-made sources with a wide range of physical and optical properties. Because of the complexity of the processes and composition, and highly inhomogeneous distribution of aerosols, accurately modeling atmospheric aerosols and their effects remains a challenge. Models have to take into account not only the aerosol and precursor emissions, but also the chemical transformation, transport, and removal processes (e.g. dry and wet depositions) to simulate the aerosol mass concentrations. Furthermore, aerosol particle size can grow in the atmosphere because the ambient water vapor can condense on the aerosol particles. This "swelling" process, called hygroscopic growth, is most commonly parameterized in the models as a function of relative humidity.

3.2.1. Estimates of Emissions

Aerosols have various sources from both natural and anthropogenic processes. Natural emissions include wind-blown mineral dust, aerosol and precursor gases from volcanic eruptions, natural wild fires, vegetation, and oceans. Anthropogenic sources include emissions from fossil fuel and biofuel combustion, industrial processes, agriculture practices, and human-induced biomass burning.

Following earlier attempts to quantify man-made primary emissions of aerosols (Turco et al., 1983; Penner et al., 1993) systematic work was undertaken in the late 1990s to calculate

Because of the complexity of the processes and composition, and highly inhomogeneous distribution of aerosols, accurately modeling atmospheric aerosols and their effects remains a challenge.

emissions of black carbon (BC) and organic carbon (OC), using fuel-use data and measured emission factors (Liousse et al., 1996; Cooke and Wilson, 1996; Cooke et al., 1999). The work was extended in greater detail and with improved attention to source-specific emission factors in Bond et al. (2004), which provides global inventories of BC and OC for the year 1996, with regional and source-category discrimination that includes contributions from industrial, transportation, residential solid-fuel combustion, vegetation and open biomass burning (forest fires, agricultural waste burning, etc.), and diesel vehicles.

Emissions from natural sources—which include wind-blown mineral dust, wildfires, sea salt, and volcanic eruptions—are less well quantified, mainly because of the difficulties of measuring emission rates in the field and the unpredictable nature of the events. Often, emissions must be inferred from ambient observations at some distance from the actual source. As an example, it was concluded (Lewis and Schwartz, 2004) that available information on size-dependent sea salt production rates could only provide order-of-magnitude estimates. The natural emissions in general can vary dramatically over space and time.

Aerosols can be produced from trace gases in the atmospheric via chemical reactions, and those aerosols are called *secondary* aerosols, as distinct from *primary* aerosols that are directly emitted to the atmosphere as aerosol particles. For example, most sulfate and nitrate aerosols are secondary aerosols that are formed from their precursor gases, sulfur dioxide (SO_2) and nitrogen oxides (NO and NO_2, collectively called NO_x), respectively. Those sources have been studied for many years and are relatively well known. By contrast, the sources of secondary organic aerosols (SOA) are poorly understood, including emissions of their precursor gases (called volatile organic compounds, VOC) from both natural and anthropogenic sources and the atmospheric production processes.

Globally, sea salt and mineral dust dominate the total aerosol mass emissions because of the large source areas and/or large particle sizes. However, sea salt and dust also have shorter atmospheric lifetimes because of their large

particle size, and are radiatively less active than aerosols with small particle size, such as sulfate, nitrate, BC, and particulate organic matter (POM, which includes both carbon and non-carbon mass in the organic aerosol, see Glossary), most of which are anthropogenic in origin.

Because the anthropogenic aerosol RF is usually evaluated (e.g., by the IPCC) as the anthropogenic perturbation since the pre-industrial period, it is necessary to estimate the historical emission trends, especially the emissions in the pre-industrial era. Compared to estimates of present-day emissions, estimates of historical emission have much larger uncertainties. Information for past years on the source types and strengths and even locations are difficult to obtain, so historical inventories from pre-industrial times to the present have to be based on limited knowledge and data. Several studies on historical emission inventories of BC and OC (e.g., Novakov et al., 2003; Ito and Penner 2005; Bond et al., 2007; Fernandes et al., 2007; Junker and Liousse, 2008), SO_2 (Stern, 2005), and various species (van Aardenne et al., 2001; Dentener et al., 2006) are available in the literature; there are some similarities and some differences among them, but the emission estimates for early times do not have the rigor of the studies for present-day emissions. One major conclusion from all these studies is that the growth of primary aerosol emissions in the 20th century was not nearly as rapid as the growth in CO_2 emissions. This is because in the late 19th and early 20th centuries, particle emissions such as BC and POM were relatively high due to the heavy use of biofuels and the lack of particulate controls on coal-burning facilities; however, as economic development continued, traditional biofuel use remained fairly constant and particulate emissions from coal burning were reduced by the application of technological controls (Bond et al., 2007). Thus, particle emissions in the 20th century did not grow as fast as CO_2 emissions, as the latter are roughly proportional to total fuel use—oil and gas included. Another challenge is estimating historical biomass burning emissions. A recent study suggested about a 40% increase in carbon emissions from biomass burning from the beginning to the end of last century (Mouillot et al., 2006), but it is difficult to verify.

Aerosols have various sources from both natural and anthropogenic processes, including dust, volcanic eruptions, fires, fossil fuel and biofuel combustion, and agricultural practices.

Table 3.1. Anthropogenic emissions of aerosols and precursors for 2000 and 1750. Adapted from Dentener et al., 2006.

Source	Species*	Emission# 2000 (Tg/yr)	Emission 1750 (Tg/yr)
Biomass burning	BC POM S	3.1 34.7 4.1	1.03 12.8 1.46
Biofuel	BC POM S	1.6 9.1 9.6	0.39 1.56 0.12
Fossil fuel	BC POM S	3.0 3.2 98.9	

\# Data source for 2000 emission: biomass burning – Global Fire Emission Dataset (GFED); biofuel BC and POM – Speciated Pollutant Emission Wizard (SPEW); biofuel sulfur – International Institute for Applied System Analysis (IIASA); fossil fuel BC and POM – SPEW; fossil fuel sulfur – Emission Database for Global Atmospheric Research (EDGAR) and IIASA. Fossil fuel emission of sulfur (S) is the sum of emission from industry, power plants, and transportation listed in Dentener et al., 2006.
* S=sulfur, including SO_2 and particulate sulfate. Most emitted as SO_2, and 2.5% emitted as sulfate.

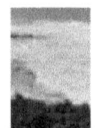

As an example, Table 3.1 shows estimated anthropogenic emissions of sulfur, BC and POM in the present day (year 2000) and pre-industrial time (1750) compiled by Dentener et al., 2006. These estimates have been used in the Aerosol Comparisons between Observations and Models (AeroCom) project (Experiment B, which uses the year 2000 emission; and Experiment PRE, which uses pre-industrial emissions), for simulating atmospheric aerosols and anthropogenic aerosol RF. The AeroCom results are discussed in Sections 3.2.2 and 3.3.

3.2.2. AEROSOL MASS LOADING AND OPTICAL DEPTH

In the global models, aerosols are usually simulated in the successive steps of sources (emission and chemical formation), transport (from source location to other area), and removal processes (dry deposition, in which particles fall onto the surface, and wet deposition by rain) that control the aerosol lifetime. Collectively, emission, transport, and removal determine the amount (mass) of aerosols in the atmosphere.

Aerosol optical depth (AOD), which is a measure of solar or thermal radiation being attenuated by aerosol particles via scattering or absorption, can be related to the atmospheric aerosol mass loading as follows:

$$AOD = MEE \cdot M \qquad (3.1)$$

where M is the aerosol mass loading per unit area (g m^{-2}), MEE is the mass extinction efficiency or specific extinction in unit of m^2 g^{-1}, which is

$$MEE = \frac{3Q_{ext}}{4\pi\rho r_{eff}} \cdot f \qquad (3.2)$$

where Q_{ext} is the extinction coefficient (a function of particle size distribution and refractive index), r_{eff} is the aerosol particle effective radius, ρ is the aerosol particle density, and f is the ratio of ambient aerosol mass (wet) to dry aerosol mass M. Here, M is the result from model-simulated atmospheric processes and MEE embodies the aerosol physical (including microphysical) and optical properties. Since Q_{ext} varies with radiation wavelength, so do MEE and AOD. AOD is the quantity that is most commonly obtained from remote sensing measurements and is frequently used for model evaluation (see Chapter 2). AOD is also a key parameter determining aerosol radiative effects.

Here the results from the recent multiple-global-model studies by the AeroCom project are summarized, as they represent the current assessment of model-simulated atmospheric aerosol loading, optical properties, and RF for the present-day. AeroCom aims to document differences in global aerosol models and compare the model output to observations. Sixteen global models participated in the AeroCom Experiment A (AeroCom-A), for which every model used their own configuration, including their own choice of estimating emissions (Kinne et al., 2006; Textor et al., 2006). Five major aerosol types: sulfate, BC, POM, dust, and sea salt, were included in the experiments, although some models had additional aerosol species. Of those major aerosol types, dust and sea-salt are predominantly natural in origin, whereas sulfate, BC, and POM have major anthropogenic sources.

Table 3.2 summarizes the model results from the AeroCom-A for several key parameters:

Table 3.2. Summary of statistics of AeroCom Experiment A results from 16 global models. Data from Textor et al. (2006) and Kinne et al. (2006), and AeroCom website (http://nansen.ipsl.jussieu.fr/AEROCOM/data.html).

Quantity	Mean	Median	Range	Stddev /mean*
Sources (Tg yr^{-1})				
Sulfate	179	186	98-232	22%
Black carbon	11.9	11.3	7.8-19.4	23%
Organic matter	96.6	96.0	53-138	26%
Dust	1840	1640	672-4040	49%
Sea salt	16600	6280	2180-121000	199%
Removal rate (day^{-1})				
Sulfate	0.25	0.24	0.19-0.39	18%
Black carbon	0.15	0.15	0.066-0.19	21%
Organic matter	0.16	0.16	0.09-0.23	24%
Dust	0.31	0.25	0.14-0.79	62%
Sea salt	5.07	2.50	0.95-35.0	188%
Lifetime (day)				
Sulfate	4.12	4.13	2.6-5.4	18%
Black carbon	7.12	6.54	5.3-15	33%
Organic matter	6.54	6.16	4.3-11	27%
Dust	4.14	4.04	1.3-7.0	43%
Sea salt	0.48	0.41	0.03-1.1	58%
Mass loading (Tg)				
Sulfate	1.99	1.98	0.92-2.70	25%
Black carbon	0.24	0.21	0.046-0.51	42%
Organic matter	1.70	1.76	0.46-2.56	27%
Dust	19.2	20.5	4.5-29.5	40%
Sea salt	7.52	6.37	2.5-13.2	54%
MEE at 550 nm (m^2 g^{-1})				
Sulfate	11.3	9.5	4.2-28.3	56%
Black carbon	9.4	9.2	5.3-18.9	36%
Organic matter	5.7	5.7	3.7-9.1	26%
Dust	0.99	0.95	0.46-2.05	45%
Sea salt	3.0	3.1	0.97-7.5	55%
AOD at 550 nm				
Sulfate	0.035	0.034	0.015-0.051	33%
Black carbon	0.004	0.004	0.002-0.009	46%
Organic matter	0.018	0.019	0.006-0.030	36%
Dust	0.032	0.033	0.012-0.054	44%
Sea salt	0.033	0.030	0.02-0.067	42%
Total AOT at 550 nm	0.124	0.127	0.065-0.151	18%

* Stddev/mean was used as the term "diversity" in Textor et al., 2006.

Sources (emission and chemical transformation), mass loading, lifetime, removal rates, and MEE and AOD at a commonly used, mid-visible, wavelength of 550 nanometer (nm). These are the globally averaged values for the year 2000. Major features and conclusions are:

- Globally, aerosol source (in mass) is dominated by sea salt, followed by dust, sulfate, POM, and BC. Over the non-desert land area, human activity is the major source of sulfate, black carbon, and organic aerosols.

- Aerosols are removed from the atmosphere by wet and dry deposition. Although sea salt dominates the emissions, it is quickly removed from the atmosphere because of its large particle size and near-surface distributions, thus having the shortest lifetime. The median lifetime of sea salt from the AeroCom-A models is less than half a day,

whereas dust and sulfate have similar lifetimes of 4 days and BC and POM 6-7 days.

- Globally, small-particle-sized sulfate, BC, and POM make up a little over 10% of total aerosol mass in the atmosphere. However, they are mainly from anthropogenic activity, so the highest concentrations are in the most populated regions, where their effects on climate and air quality are major concerns.

- Sulfate and BC have their highest MEE at mid-visible wavelengths, whereas dust is lowest among the aerosol types modeled. That means for the same amount of aerosol mass, sulfate and BC are more effective at attenuating (scattering or absorbing) solar radiation than dust. This is why the sulfate AOD is about the same as dust AOD even though the atmospheric amount of sulfate mass is 10 times less than that of the dust.

- There are large differences, or diversities, among the models for all the parameters listed in Table 3.2. The largest model diversity, shown as the % standard deviation from the all-model-mean and the range (minimum and maximum values) in Table 3.2, is in sea salt emission and removal; this is mainly associated with the differences in particle size range and source parameterizations in each model. The diversity of sea salt atmospheric loading however is much smaller than that of sources or sinks, because the largest particles have the shortest lifetimes even though they comprise the largest fraction of emitted and deposited mass.

- Among the key parameters compared in Table 3.2, the models agree best for simulated total AOD – the % of standard deviation from the model mean is 18%, with the extreme values just a factor of 2 apart. The median value of the multi-model simulated global annual mean total AOD, 0.127, is also in agreement with the global mean values from recent satellite measurements. However, despite the general agreement in total AOD, there are significant diversities at the individual component level for aerosol optical thickness, mass loading, and mass extinction efficiency. This indicates that uncertainties in assessing aerosol climate forcing are still large, and they depend not only on total AOD but also on aerosol absorption and scattering direction (called asymmetry factor; see next page and Glossary), both of which are determined by aerosol physical

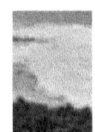

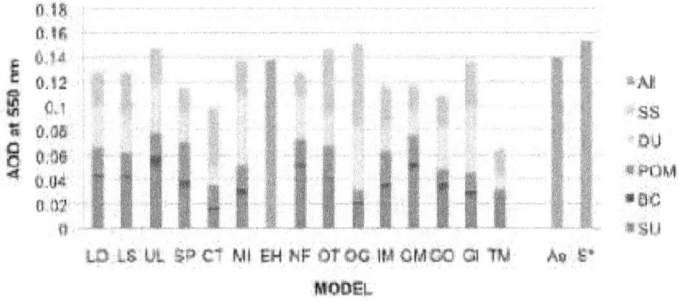

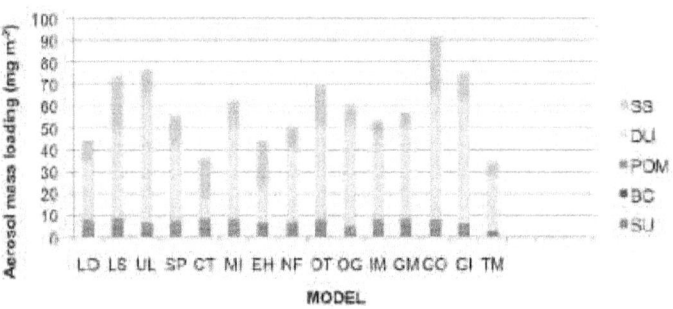

Figure 3.1. Global annual averaged AOD (upper panel) and aerosol mass loading (lower panel) with their components simulated by 15 models in AeroCom-A (excluding one model which only reported mass). SU=sulfate, BC=black carbon, POM=particulate organic carbon, DU=dust, SS=sea salt. Model abbreviations: LO=LOA (Lille, Fra), LS=LSCE (Paris, Fra), UL=ULAQ (L'Aquila, Ita), SP=SPRINTARS (Kyushu, Jap), CT=ARQM (Toronto, Can), MI=MIRAGE (Richland, USA), EH=ECHAM5 (MPI-Hamburg, Ger), NF=CCM-Match (NCAR-Boulder, USA), OT=Oslo-CTM (Oslo, Nor), OG=OLSO-GCM (Oslo, Nor) [prescribed background for DU and SS], IM=IMPACT (Michigan, USA), GM=GFDL-Mozart (Princeton, NJ, USA), GO=GOCART (NASA-GSFC, Washington DC, USA), GI=GISS (NASA-GISS, New York, USA), TM=TM5 (Utrecht, Net). Also shown in the upper panel are the averaged observation data from AERONET (Ae) and the satellite composite (S*). See Kinne et al. (2006) for details. Figure produced from data in Kinne et al. (2006).

and optical properties. In addition, even with large differences in mass loading and MEE among different models, these terms could compensate for each other (eq. 3.1) to produce similar AOD. This is illustrated in Figure 3.1. For example, model LO and LS have quite different mass loading (44 and 74 mg m-2, respectively), especially for dust and sea salt amount, but they produce nearly identical total AOD (0.127 and 0.128, respectively).

• Because of the large spatial and temporal variations of aerosol distributions, regional and seasonal diversities are even larger than the diversity for global annual means.

To further isolate the impact of the differences in emissions on the diversity of simulated aerosol mass loading, identical emissions for aerosols and their precursor were used in the AeroCom Experiment B exercise in which 12 of the 16 AeroCom-A models participated (Textor et al., 2007). The comparison of the results and diversity between AeroCom-A and -B for the same models showed that using harmonized emissions does not significantly reduce model diversity for the simulated global mass and AOD fields, indicating that the differences in atmospheric processes, such as transport, removal, chemistry, and aerosol microphysics, play more important roles than emission in creating diversity among the models. This outcome is somewhat different from another recent study, in which the differences in calculated clear-sky aerosol RF between two models (a regional model STEM and a global model MOZART) were attributed mostly to the differences in emissions (Bates et al., 2006), although the conclusion was based on only two model simulations for a few focused regions. It is highly recommended from the outcome of AeroCom-A and -B that, although more detailed evaluation for each individual process is needed, multi-model ensemble results, e.g., median values of multi-model output variables, should be used to estimate aerosol RF, due to their greater robustness, relative to individual models, when compared to observations (Textor et al., 2006, 2007; Schulz et al., 2006).

3.3. Calculating Aerosol Direct Radiative Forcing

The three parameters that define the aerosol direct RF are the AOD, the single scattering

albedo (SSA), and the asymmetry factor (g), all of which are wavelength dependent. AOD is indicative of how much aerosol exists in the column, SSA is the fraction of radiation being scattered versus the total attenuation (scattered and absorbed), and the g relates to the direction of scattering that is related to the size of the particles (see Chapter 1). An indication of the particle size is provided by another parameter, the Ångström exponent ($\mathring{A}$), which is a measure of differences of AOD at different wavelengths. For typical tropospheric aerosols, $\mathring{A}$ tends to be inversely dependent on particle size; larger values of $\mathring{A}$ are generally associated with smaller aerosols particles. These parameters are further related; for example, for a given composition, the ability of a particle to scatter radiation decreases more rapidly with decreasing size than does its ability to absorb, so at a given wavelength varying $\mathring{A}$ can change SSA. Note that AOD, SSA, g, $\mathring{A}$, and all the other parameters in eq. 3.1 and 3.2 vary with space and time due to variations of both aerosol composition and relative humidity, which influence these characteristics.

In the recent AeroCom project, aerosol direct RF for the solar spectral wavelengths (or shortwave) was assessed based on the 9 models that participated in both Experiment B and PRE in which identical, prescribed emissions for present (year 2000) and pre-industrial time (year 1750) listed in Table 3.1 were used across the models (Schulz et al., 2006). The anthropogenic direct RF was obtained by subtracting Aero-Com-PRE from AeroCom-B simulated results. Because dust and sea salt are predominantly from natural sources, they were not included in the anthropogenic RF assessment although the land use practice can contribute to dust emissions as "anthropogenic". Other aerosols that were not considered in the AeroCom forcing assessment were natural sulfate (e.g. from volcanoes or ocean) and POM (e.g. from biogenic hydrocarbon oxidation), as well as nitrate. The aerosol direct forcing in the AeroCom assessment thus comprises three major anthropogenic aerosol components sulfate, BC, and POM.

The IPCC AR4 (IPCC, 2007) assessed anthropogenic aerosol RF based on the model results published after the IPCC TAR in 2001, including those from the AeroCom study discussed above. These results (adopted from IPCC

There are large differences in simulated aerosol fields among the models.

AR4) are shown in Table 3.3 for sulfate and Table 3.4 for carbonaceous aerosols (BC and POM), respectively. All values listed in Table 3.3 and 3.4 refer to anthropogenic perturbation, i.e., excluding the natural fraction of these aerosols. In addition to the mass burden, MEE, and AOD, Table 3.3 and 3.4 also list the "normalized forcing", also known as "forcing efficiency", one for the forcing per unit AOD, and the other the forcing per gram of aerosol mass (dry). For some models, aerosols are externally mixed, that is, each aerosol particle contains

Table 3.3. Sulfate mass loading, MEE and AOD at 550 nm, shortwave radiative forcing at the top of the atmosphere, and normalized forcing with respect to AOD and mass. All values refer to anthropogenic perturbation. Adapted from IPCC AR4 (2007) and Schulz et al. (2006).

Model	Mass load (mg m⁻²)	MEE (m² g⁻¹)	AOD at 550 nm	TOA Forcing (W m⁻²)	Forcing/AOD (W m⁻²)	Forcing/mass (W g⁻¹)
Published since IPCC 2001						
A CCM3	2.23			-0.56		-251
B GEOSCHEM	1.53	11.8	0.018	-0.33	-18	-216
C GISS	3.30	6.7	0.022	-0.65	-30	-197
D GISS	3.27			-0.96		-294
E GISS*	2.12			-0.57		-269
F SPRINTARS	1.55	9.7	0.015	-0.21		-135
G LMD	2.76			-0.42		-152
H LOA	3.03	9.9	0.03	-0.41	-14	-135
I GATORG	3.06			-0.32		-105
J PNNL	5.50	7.6	0.042	-0.44	-10	-80
K UIO-CTM	1.79	10.6	0.019	-0.37	-19	-207
L UIO-GCM	2.28			-0.29		-127
AeroCom: Identical emissions used for year 2000 and 1750						
M UMI	2.64	7.6	0.02	-0.58	-29	-220
N UIO-CTM	1.70	11.2	0.019	-0.36	-19	-212
O LOA	3.64	9.6	0.035	-0.49	-14	-135
P LSCE	3.01	7.6	0.023	-0.42	-18	-140
Q ECHAM5-HAM	2.47	6.5	0.016	-0.46	-29	-186
R GISS**	1.34	4.5	0.006	-0.19	-32	-142
S UIO-GCM	1.72	7.0	0.012	-0.25	-21	-145
T SPRINTARS	1.19	10.9	0.013	-0.16	-12	-134
U ULAQ	1.62	12.3	0.02	-0.22	-11	-136
Average A-L	**2.70**	**9.4**	**0.024**	**-0.46**	**-18**	**-181**
Average M-U	**2.15**	**8.6**	**0.018**	**-0.35**	**-21**	**-161**
Minimum A-U	**1.19**	**4.5**	**0.006**	**-0.96**	**-32**	**-294**
Maximum A-U	**5.50**	**12.3**	**0.042**	**-0.16**	**-10**	**-80**
Std dev A-L	**1.09**	**1.9**	**0.010**	**0.202**	**7**	**68**
Std dev M-U	**0.83**	**2.6**	**0.008**	**0.149**	**8**	**35**
%Stddev/avg A-L	**40%**	**20%**	**41%**	**44%**	**38%**	**38%**
%Stddev/avg M-U	**39%**	**30%**	**45%**	**43%**	**37%**	**22%**

Model abbreviations: CCM3=Community Climate Model; GEOSCHEM=Goddard Earth Observing System-Chemistry; GISS=Goddard Institute for Space Studies; SPRINTARS=Spectral Radiation-Transport Model for Aerosol Species; LMD=Laboratoire de Meteorologie Dynamique; LOA=Laboratoire d'Optique Atmospherique; GATORG=Gas, Aerosol Transport and General circulation model; PNNL=Pacific Northwest National Laboratory; UIO-CTM=Univerity of Oslo CTM; UIO-GCM=University of Oslo GCM; UMI=University of Michigan; LSCE=Laboratoire des Sciences du Climat et de l'Environment; ECHAMS5-HAM=European Centre Hamburg with Hamburg Aerosol Module; ULAQ=University of IL'Aquila.

only one aerosol type such as sulfate, whereas other models allow aerosols to mix internally to different degrees, that is, each aerosol particle can have more than one component, such as black carbon coated with sulfate. For models with internal mixing of aerosols, the compo-

nent values for AOD, MEE, and forcing were extracted (Schulz et al., 2006).

Considerable variation exists among these models for all quantities in Table 3.3 and 3.4. The RF for all the components varies by a factor of

Table 3.4. Particulate organic matter (POM) and black carbon (BC) mass loading, AOD at 550 nm, shortwave radiative forcing at the top of the atmosphere, and normalized forcing with respect to AOD and mass. All values refer to anthropogenic perturbation. Based on IPCC AR4 (2007) and Schulz et al. (2006).

| Model | POM | | | | | | BC | | | | | |
	Mass load (mg m⁻²)	MEE (m² g⁻¹)	AOD at 550 nm	TOA Forcing (W m⁻²)	Forcing/ AOD (W m⁻²)	Forcing/ mass (W g⁻¹)	Mass load (mg m⁻²)	MEE (m² g⁻¹)	AOD at 550 nm x1000	TOA Forcing (W m⁻²)	Forcing/ AOD (W m⁻²)	Forcing/ mass (W g⁻¹)
Published since IPCC 2001												
A SPRINTARS				-0.24		-107				0.36		
B LOA	2.33	6.9	0.016	-0.25	-16	-140	0.37			0.55		
C GISS	1.86	9.1	0.017	-0.26	-15	-161	0.29			0.61		
D GISS	1.86	8.1	0.015	-0.30	-20	-75	0.29			0.35		
E GISS*	2.39			-0.18		-92	0.39			0.50		
F GISS	2.49			-0.23		-101	0.43			0.53		
G SPRINTARS	2.67	10.9	0.029	-0.27	-9	-23	0.53			0.42		
H GATORG	2.56			-0.06		-112	0.39			0.55		
I MOZGN	3.03	5.9	0.018	-0.34	-19							
J CCM							0.33			0.34		
K UIO-GCM							0.30			0.19		
AeroCom: Identical emissions for year 2000 & 1750												
L UMI	1.16	5.2	0.0060	-0.23	-38	-198	0.19	6.8	1.29	0.25	194	1316
M UIO-CTM	1.12	5.2	0.0058	-0.16	-28	-143	0.19	7.1	1.34	0.22	164	1158
N LOA	1.41	6.0	0.0085	-0.16	-19	-113	0.25	7.9	1.98	0.32	162	1280
O LSCE	1.50	5.3	0.0079	-0.17	-22	-113	0.25	4.4	1.11	0.30	270	1200
P ECHAM5-HAM	1.00	7.7	0.0077	-0.10	-13	-100	0.16	7.7	1.23	0.20	163	1250
Q GISS**	1.22	4.9	0.0060	-0.14	-23	-115	0.24	7.6	1.83	0.22	120	917
R UIO-GCM	0.88	5.2	0.0046	-0.06	-13	-68	0.19	10.3	1.95	0.36	185	1895
S SPRINTARS	1.84	10.9	0.0200	-0.10	-5	-54	0.37	9.5	3.50	0.32	91	865
T ULAQ	1.71	4.4	0.0075	-0.09	-12	-53	0.38	7.6	2.90	0.08	28	211
Average A-K	2.40	8.2	0.019	-0.24	-16	-102	0.37			0.44		1242
Average L-T	1.32	6.1	0.008	-0.13	-19	-106	0.25	7.7	1.90	0.25	153	1121
Minimum A-T	0.88	4.4	0.005	-0.34	-38	-198	0.16	4.4	1.11	0.08	28	211
Maximum A-T	3.03	10.9	0.029	-0.06	-5	-23	0.53	10.3	3.50	0.61	270	2103
Std dev A-K	0.39	1.7	0.006	0.09	4	41	0.08			0.06		384
Std dev L-T	0.32	2.0	0.005	0.05	10	46	0.08	1.6	0.82	0.09	68	450
%Stddev/avg A-K	16%	21%	30%	36%	26%	41%	22%			23%		31%
%Stddev/avg L-T	25%	33%	56%	39%	52%	43%	32%	21%	43%	34%	45%	40%

Although black carbon has the lowest mass loading and optical depth, it is the only aerosol species that absorbs strongly, causing positive forcing to warm the atmosphere, in contrast to other aerosols that impose negative forcing to cool the atmosphere.

6 or more: Sulfate from 0.16 to 0.96 W m^{-2}, POM from -0.06 to -0.34 W m^{-2}, and BC from +0.08 to +0.61 W m^{-2}, with the standard deviation in the range of 30 to 40% of the ensemble mean. It should be noted that although BC has the lowest mass loading and AOD, it is the only aerosol species that absorbs strongly, causing positive forcing to warm the atmosphere, in contrast to other aerosols that impose negative forcing to cool the atmosphere. As a result, the net anthropogenic aerosol forcing as a whole becomes less negative when BC is included. The global average anthropogenic aerosol direct RF at the top of the atmosphere (TOA) from the models, together with observation-based estimates (see Chapter 2), is presented in Figure 3.2. Note the wide range for forcing in Figure 3.2. The comparison with observation-based estimates shows that the model estimated forcing is in general lower, partially because the forcing value from the model is the difference between present-day and pre-industrial time, whereas the observation-derived quantity is the difference between an atmosphere with and without anthropogenic aerosols, so the "background" value that is subtracted from the total forcing is higher in the models.

The discussion so far has dealt with global average values. The geographic distributions of multi-model aerosol direct RF has been evaluated among the AeroCom models, which are shown in Figure 3.3 for total and anthropogenic AOD at 550 nm and anthropogenic aerosol RF at TOA, within the atmospheric column, and at the surface. Globally, anthropogenic AOD is about 25% of total AOD (Figure 3.3a and b) but is more concentrated over polluted regions in Asia, Europe, and North America and biomass burning regions in tropical southern Africa and South America. At TOA, anthropogenic aerosol causes negative forcing over mid-latitude continents and oceans with the most negative values (-1 to -2 W m^{-2}) over polluted regions (Figure 3.3c). Although anthropogenic aerosol has a cooling effect at the surface with surface forcing values down to -10 W m^{-2} over China, India, and tropical Africa (Figure 3.3e), it warms the atmospheric column with the largest effects again over the polluted and biomass burning regions. This heating effect will change the atmospheric circulation and can affect the weather and precipitation (e.g., Kim et al., 2006).

Basic conclusions from forward modeling of aerosol direct RF are:

- The most recent estimate of all-sky shortwave aerosol direct RF at TOA from anthropogenic sulfate, BC, and POM (mostly from fossil fuel/biofuel combustion and biomass burning) is -0.22 ± 0.18 W m^{-2} averaged globally, exerting a net cooling effect. This value would represent the low-end of the forcing magnitude, since some potentially significant anthropogenic aerosols, such as nitrate and dust from human activities are not included because of their highly uncertain sources and processes. IPCC AR4 had adjusted the total anthropogenic aerosol direct RF to -0.5 ± 0.4 W m^{-2} by adding estimated anthropogenic nitrate and dust forcing values based on limited modeling studies and by considering the observation-based estimates (see Chapter 2).

- Both sulfate and POM negative forcing whereas BC causes positive forcing because of its highly absorbing nature. Although BC comprises only a small fraction of anthropogenic aerosol mass load and AOD, its forcing efficiency (with respect to either AOD or mass) is an order of magnitude stronger than sulfate and POM, so its positive shortwave forcing largely offsets the negative forcing

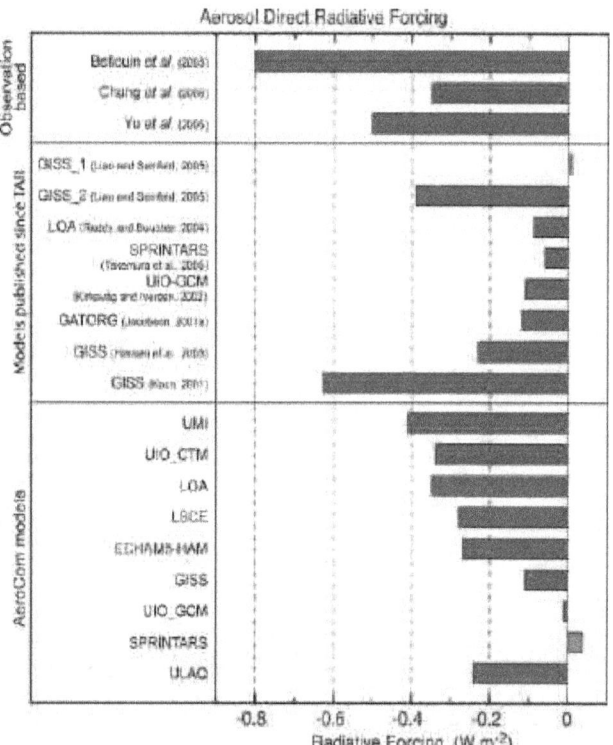

Figure 3.2. Aerosol direct radiative forcing in various climate and aerosol models. Observed values are shown in the top section. From IPCC (2007).

(a) Mean AOD 550 nm

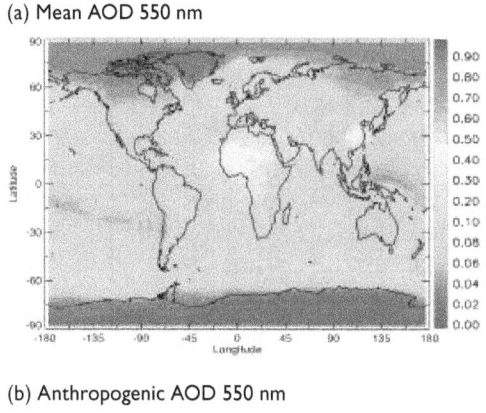

(c) Anthro. aerosol TOA forcing (W m-2)

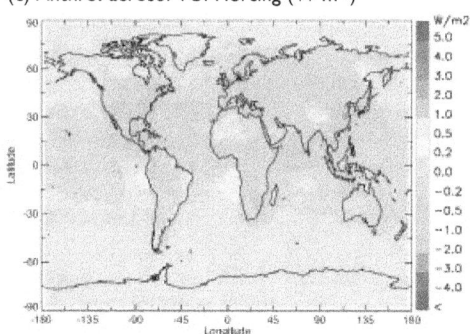

(b) Anthropogenic AOD 550 nm

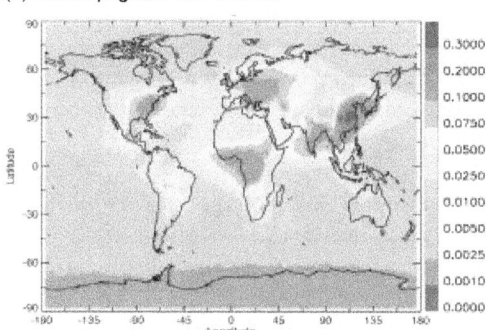

(d) Anthro. aerosol atmospheric forcing (W m-2)

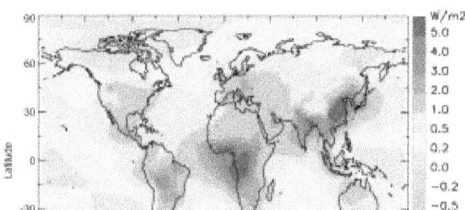

(e) Anthro. Aerosol surface forcing (W m-2)

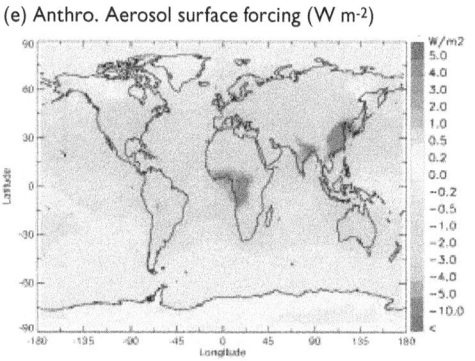

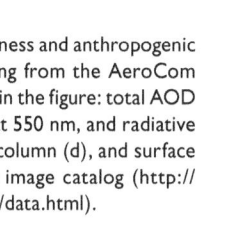

Figure 3.3. Aerosol optical thickness and anthropogenic shortwave all-sky radiative forcing from the AeroCom study (Schulz et al., 2006). Shown in the figure: total AOD (a) and anthropogenic AOD (b) at 550 nm, and radiative forcing at TOA (c), atmospheric column (d), and surface (e). Figures from the AeroCom image catalog (http://nansen.ipsl.jussieu.fr/AEROCOM/data.html).

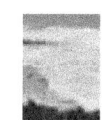

from sulfate and POM. This points out the importance of improving the model ability to simulate each individual aerosol components more accurately, especially black carbon. Separately, it is estimated from recent model studies that anthropogenic sulfate, POM, and BC forcings at TOA are -0.4, -0.18, +0.35 W m-2, respectively. The anthropogenic nitrate and dust forcings are estimated at -0.1 W m-2 for each, with uncertainties exceeds 100% (IPCC AR4, 2007).

- In contrast to long-lived greenhouse gases, anthropogenic aerosol RF exhibits significant regional and seasonal variations. The forcing magnitude is the largest over the industrial and biomass burning source regions, where the magnitude of the negative aerosol forcing can be of the same magnitude or even stronger than that of positive greenhouse gas forcing.

- There is a large spread of model-calculated aerosol RF even in the global annual averaged values. The AeroCom study shows that the model diversity at some locations (mostly East Asia and African biomass burning regions) can reach ±3 W m-2, which is an order of magnitude above the global averaged forcing value of -0.22 W m-2. The large diversity reflects the low level of current understanding of aerosol radiative forcing, which is compounded by uncertainties in emissions, transport, transformation, removal, particle size, and optical and microphysical (including hygroscopic) properties.

In contrast to long-lived greenhouse gases, anthropogenic aerosol radiative forcing exhibits significant regional and seasonal variations.

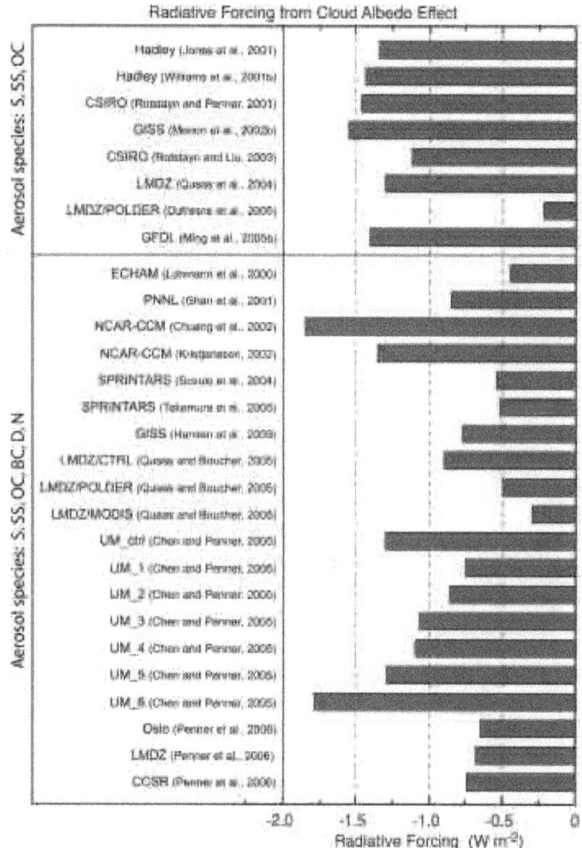

Fig. 3.4. Radiative forcing from the cloud albedo effect (1st aerosol indirect effect) in the global climate models used in IPCC 2007 (IPCC Fig. 2.14). For additional model designations and references, see IPCC 2007, chapter 2. Species included in the lower panel are sulfate, sea salt, organic and black carbon, dust and nitrates; in the top panel, only sulfate, sea salt and organic carbon are included.

centrations, therefore, may increase the ambient concentrations of CCN and IN, affecting cloud properties. For a fixed cloud liquid water content, a CCN increase will lead to more cloud droplets so that the cloud droplet size will decrease. That effect leads to brighter clouds, the enhanced albedo then being referred to as the "cloud albedo effect" (Twomey, 1977), also known as the first indirect effect. If the droplet size is smaller, it may take longer to rainout, leading to an increase in cloud lifetime, hence the "cloud lifetime" effect (Albrecht, 1989), also called the second indirect effect. Approximately one-third of the models used for the IPCC 20th century climate change simulations incorporated an aerosol indirect effect, generally (though not exclusively) considered only with sulfates.

Shown in Figure 3.4 are results from published model studies indicating the different RF values from the cloud albedo effect. The cloud albedo effect ranges from -0.22 to -1.85 W m-2; the lowest estimates are from simulations that constrained representation of aerosol effects on clouds with satellite measurements of drop size vs. aerosol index. In view of the difficulty of quantifying this effect remotely (discussed later), it is not clear whether this constraint provides an improved estimate. The estimate in the IPCC AR4 ranges from +0.4 to -1.1 W m-2, with a "best-guess" estimate of 0.7 W m-2.

The representation of cloud effects in GCMs is considered below. However, it is becoming increasingly clear from studies based on high resolution simulations of aerosol-cloud interactions that there is a great deal of complexity that is unresolved in climate models. This point is examined again in section 3.4.4.

Most models did not incorporate the "cloud lifetime effect". Hansen et al. (2005) compared this latter influence (in the form of time-averaged cloud area or cloud cover increase) with the cloud albedo effect. In contrast to the discussion in IPCC (2007), they argue that the cloud cover effect is more likely to be the dominant one, as suggested both by cloud-resolving model studies (Ackerman et al., 2004) and satellite observations (Kaufman et al., 2005c). The cloud albedo effect may be partly offset by reduced cloud thickness accompanying aerosol pollutants, producing a meteorological (cloud)

Anthropogenic aerosols cool the surface but heat the atmosphere.

- In spite of the relatively small value of forcing at TOA, the magnitudes of anthropogenic forcing at the surface and within the atmospheric column are considerably larger: -1 to -2 W m-2 at the surface and +0.8 to +2 W m-2 in the atmosphere. Anthropogenic aerosols thus cool the surface but heat the atmosphere, on average. Regionally, the atmospheric heating can reach annually averaged values exceeding 5 W m-2 (Figure 3.3d). These regional effects and the negative surface forcing are expected to exert an important effect on climate through alteration of the hydrological cycle.

3.4. Calculating Aerosol Indirect Forcing
3.4.1. Aerosol Effects on Clouds

A subset of the aerosol particles can act as cloud condensation nuclei (CCN) and/or ice nuclei (IN). Increases in aerosol particle con-

rather than aerosol effect (see the discussion in Lohmann and Feichter, 2005). (The distinction between meteorological feedback and aerosol forcing can become quite opaque; as noted earlier, the term feedback is restricted here to those processes that are responding to a change in temperature.) Nevertheless, both aerosol indirect effects were utilized in Hansen et al. (2005), with the second indirect effect calculated by relating cloud cover to the aerosol number concentration, which in turn is a function of sulfate, nitrate, black carbon and organic carbon concentration. Only the low altitude cloud influence was modeled, principally because there are greater aerosol concentrations at low levels, and because low clouds currently exert greater cloud RF. The aerosol influence on high altitude clouds, associated with IN changes, is a relatively unexplored area for models and as well for process-level understanding.

Hansen et al. (2005) used coefficients to normalize the cooling from aerosol indirect effects to between -0.75 and -1 W m-2, based on comparisons of modeled and observed changes in the diurnal temperature range as well as some satellite observations. The response of the GISS model to the direct and two indirect effects is shown in Figure 3.5. As parameterized, the cloud lifetime effect produced somewhat

greater negative RF (cooling), but this was the result of the coefficients chosen. Geographically, it appears that the "cloud cover" effect produced slightly more cooling in the Southern Hemisphere than did the "cloud albedo" response, with the reverse being true in the Northern Hemisphere (differences on the order of a few tenths °C).

3.4.2. Model Experiments

There are many different factors that can explain the large divergence of aerosol indirect effects in models (Fig. 3.4). To explore this in more depth, Penner et al. (2006) used three general circulation models to analyze the differences between models for the first indirect effect, as well as a combined first plus second indirect effect. The models all had different cloud and/or convection parameterizations.

In the first experiment, the monthly average aerosol mass and size distribution of, effectively, sulfate aerosol were prescribed, and all models followed the same prescription for parameterizing the cloud droplet number concentration (CDNC) as a function of aerosol concentration. In that sense, the only difference among the models was their separate cloud formation and radiation schemes. The different models all produced similar droplet

There are many different factors that can explain the large divergence of indirect effects in models.

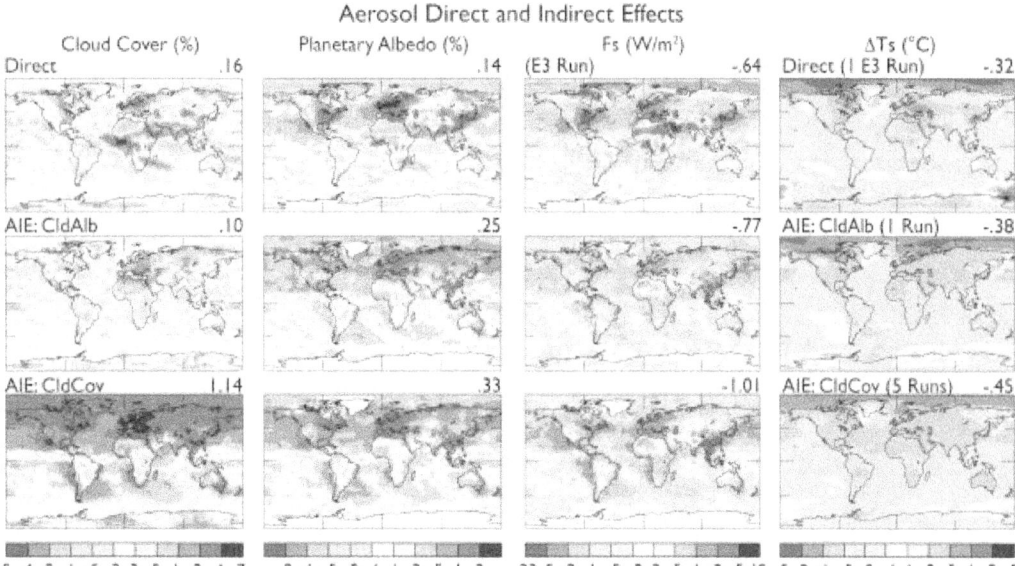

Fig. 3.5. Anthropogenic impact on cloud cover, planetary albedo, radiative flux at the surface (while holding sea surface temperatures and sea ice fixed) and surface air temperature change from the direct aerosol forcing (top row), the 1st indirect effect (second row) and the second indirect effect (third row). The temperature change is calculated from years 81-120 of a coupled atmosphere simulation with the GISS model. From Hansen et al. (2005).

effective radii, and therefore shortwave cloud forcing, and change in net outgoing whole sky radiation between pre-industrial times and the present. Hence the first indirect effect was not a strong function of the cloud or radiation scheme. The results for this and the following experiments are presented in Figure 3.6, where the experimental results are shown sequentially from left to right for the whole sky effect, and in Table 3.5 for the clear-sky and cloud forcing response as well.

The change in cloud forcing is the difference between whole sky and clear sky outgoing radiation in the present day minus pre-industrial simulation. The large differences seen between experiments 5 and 6 are due to the inclusion of the clear sky component of aerosol scattering and absorption (the direct effect) in experiment 6.

In the second experiment, the aerosol mass and size distribution were again prescribed, but now each model used its own formulation for relating aerosols to droplets. In this case one of the models produced larger effective radii and therefore a much smaller first indirect aerosol

effect (Figure 3.6, Table 3.5). However, even in the two models where the effective radius change and net global forcing were similar, the spatial patterns of cloud forcing differ, especially over the biomass burning regions of Africa and South America.

The third experiment allowed the models to relate the change in droplet size to change in precipitation efficiency (i.e., they were now also allowing the second indirect effect - smaller droplets being less efficient rain producers – as well as the first). The models utilized the same relationship for autoconversion of cloud droplets to precipitation. Changing the precipitation efficiency results in all models producing an increase in cloud liquid water path, although the effect on cloud fraction was smaller than in the previous experiments. The net result was to increase the negative radiative forcing in all three models, albeit with different magnitudes: for two of the models the net impact on outgoing shortwave radiative increased by about 20%, whereas in the third model (which had the much smaller first indirect effect), it was magnified by a factor of three.

In the fourth experiment, the models were now each allowed to use their own formulation to relate aerosols to precipitation efficiency. This introduced some additional changes in the whole sky shortwave forcing (Figure 3.6).

In the fifth experiment, models were allowed to produce their own aerosol concentrations, but were given common sources. This produced the largest changes in the RF in several of the models. Within any one model, therefore, the change in aerosol concentration has the largest effect on droplet concentrations and effective radii. This experiment too resulted in large changes in RF.

In the last experiment, the aerosol direct effect was included, based on the full range of aerosols used in each model. While the impact on the whole-sky forcing was not large, the addition of aerosol scattering and absorption primarily affected the change in clear sky radiation (Table 3.5).

The results of this study emphasize that in addition to questions concerning cloud physics, the differences in aerosol concentrations among the

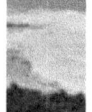

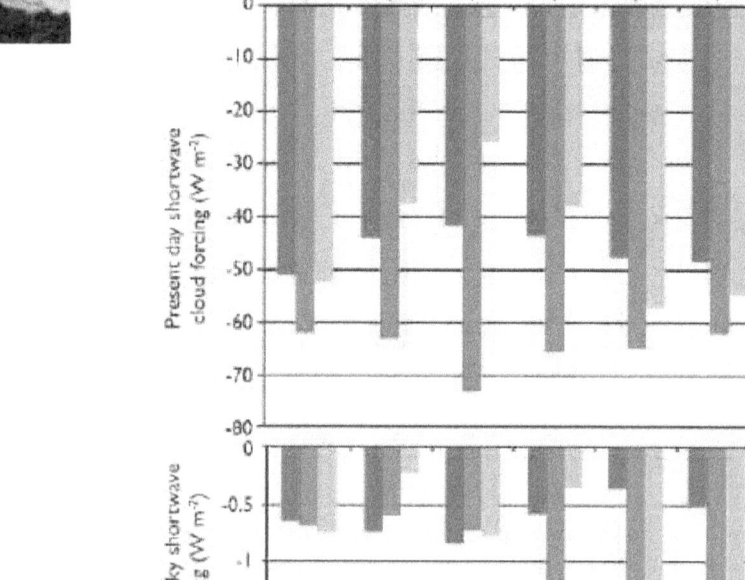

Fig. 3.6. Global average present-day short wave cloud forcing at TOA (top) and change in whole sky net outgoing shortwave radiation (bottom) between the present-day and pre-industrial simulations for each model in each experiment. Adapted from Penner et al. (2006).

Table 3.5. Differences in present day and pre-industrial outgoing solar radiation (W m⁻²) in the different experiments. Adapted from Penner et al. (2006).

MODEL	EXP 1	EXP 2	EXP 3	EXP 4	EXP 5	EXP 6
Whole-sky						
CAM-Oslo	-0.648	-0.726	-0.833	-0.580	-0.365	-0.518
LMD-Z	-0.682	-0.597	-0.722	-1.194	-1.479	-1.553
CCSR	-0.739	-0.218	-0.733	-0.350	-1.386	-1.386
Clear-sky						
CAM-Oslo	-0.063	-0.066	-0.026	0.014	-0.054	-0.575
LMD-Z	-0.054	0.019	0.030	-0.066	-0.126	-1.034
CCSR	0.018	-0.007	-0.045	-0.008	0.018	-1.160
Cloud-forcing						
CAM-Oslo	-0.548	-0.660	-0.807	-0.595	-0.311	0.056
LMD-Z	-0.628	-0.616	-0.752	-1.128	-1.353	-0.518
CCSR	-0.757	-0.212	-0.728	-0.345	-1.404	-0.200

EXP1: tests cloud formation and radiation schemes
EXP2: tests formulation for relating aerosols to droplets
EXP3: tests inclusion of droplet size influence on precipitation efficiency
EXP4: tests formulation of droplet size influence on precipitation efficiency
EXP5: tests model aerosol formulation from common sources
EXP6: added the direct aerosol effect

models play a strong role in inducing differences in the indirect effect(s), as well as the direct one.

Observational constraints on climate model simulations of the indirect effect with satellite data (e.g., MODIS) have been performed previously in a number of studies (e.g., Storelvmo et al., 2006, Lohmann et al., 2006, Quaas et al., 2006, Menon et al., 2008). These have been somewhat limited since the satellite retrieved data used do not have the vertical profiles needed to resolve aerosol and cloud fields (e.g., cloud droplet number and liquid water content); the temporal resolution of simultaneous aerosol and cloud product retrievals are usually not available at a frequency of more than one a day; and higher level clouds often obscure low clouds and aerosols. Thus, the indirect effect, especially the second indirect effect, remains, to a large extent, unconstrained by satellite observations. However, improved measurements of aerosol vertical distribution from the newer generation of sensors on the A-train platform may provide a better understanding of changes to cloud properties from aerosols. Simulating the top-of-atmosphere reflectance for comparison to satellite measured values

could be another way to compare model with observations, which would eliminate the inconsistent assumptions of aerosol optical properties and surface reflectance encountered when compared the model calculated and satellite retrieved AOD values.

3.4.3. ADDITIONAL AEROSOL INFLUENCES
Various observations have empirically related aerosols injected from biomass burning or industrial processes to reductions in rainfall (e.g., Warner, 1968; Eagan et al., 1974; Andreae et al., 2004; Rosenfeld, 2000). There are several potential mechanisms associated with this response.

In addition to the two indirect aerosol effects noted above, a process denoted as the "semidirect" effect involves the absorption of solar radiation by aerosols such as black carbon and dust. The absorption increases the temperature, thus lowering the relative humidity and producing evaporation, hence a reduction in cloud liquid water. The impact of this process depends strongly on what the effective aerosol absorption actually is; the more absorbing the aerosol, the larger the potential positive forcing on climate (by reducing low level clouds and allowing more solar radiation to reach the surface). This effect is responsible for shifting the critical value of SSA (separating aerosol cooling from aerosol warming) from 0.86 with fixed clouds to 0.91 with varying clouds (Hansen et al., 1997). Reduction in cloud cover and liquid water is one way aerosols could reduce rainfall.

More generally, aerosols can alter the location of solar radiation absorption within the system, and this aspect alone can alter climate and precipitation even without producing any change in net radiation at the top of the atmosphere (the usual metric for climate impact). By decreasing solar absorption at the surface, aerosols (from both the direct and indirect effects) reduce the energy available for evapotranspiration, potentially resulting in a decrease in precipitation. This effect has been suggested as the reason for the decrease in pan evaporation over the last 50 years (Roderick and Farquhar, 2002). The decline in solar radiation at the surface appears to have ended in the 1990s (Wild et al., 2005), perhaps because of reduced aerosol emissions in industrial areas (Kruger and Grasl, 2002), although this issue is still not settled.

Energy absorption by aerosols above the boundary layer can also inhibit precipitation by warming the air at altitude relative to the surface, i.e., increasing atmospheric stability. The increased stability can then inhibit convection, affecting both rainfall and atmospheric circulation (Ramanathan et al., 2001a; Chung and Zhang, 2004). To the extent that aerosols decrease droplet size and reduce precipitation efficiency, this effect by itself could result in lowered rainfall values locally.

In their latest simulations, Hansen et al. (2007) did find that the indirect aerosol effect reduced tropical precipitation; however, the effect is similar regardless of which of the two indirect effects is used, and also similar to the direct effect. So it is likely that the reduction of tropical precipitation is because of aerosol induced cooling at the surface and the consequent reduced evapotranspiration. Similar conclusions were reached by Yu et al. (2002) and Feingold et al. (2005). In this case, the effect is a feedback and not a forcing.

The local precipitation change, through its impacts on dynamics and soil moisture, can have

large positive feedbacks. Harvey (2004) concluded from assessing the response to aerosols in eight coupled models that the aerosol impact on precipitation was larger than on temperature. He also found that the precipitation impact differed substantially among the models, with little correlation among them.

Recent GCM simulations have further examined the aerosol effects on hydrological cycle. Ramanathan et al. (2005) showed from fully coupled ocean–atmosphere GCM experiments that the "solar dimming" effect at the surface, i.e., the reduction of solar radiation reaching the surface, due to the inclusion of absorbing aerosol forcing causes a reduction in surface evaporation, a decrease in meridional sea surface temperature (SST) gradient and an increase in atmospheric stability, and a reduction in rainfall over South Asia. Lau and Kim (2006) examined the direct effects of aerosol on the monsoon water cycle variability from GCM simulations with prescribed realistic global aerosol forcing and proposed the "elevated heat pump" effect, suggesting that atmospheric heating by absorbing aerosols (dust and black carbon), through water cycle feedback, may lead to a strengthening of the South Asia monsoon. These model results are not necessarily at odds with each other, but rather illustrate the complexity of the aerosol–monsoon interactions that are associated with different mechanisms, whose relative importance in affecting the monsoon may be strongly dependent on spatial and temporal scales and the timing of the monsoon. These results may be model dependent and should be further examined.

3.4.4. High Resolution Modeling
Largely by its nature, the representation of the interaction between aerosol and clouds in GCMs is poorly resolved. This stems in large part from the fact that GCMs do not resolve convection on their large grids (order of several hundred km), that their treatment of cloud microphysics is rather crude, and that as discussed previously, their representation of aerosol needs improvement. Superparametrization efforts (where standard cloud parameterizations in the GCM are replaced by resolving clouds in each grid column of the GCM via a cloud resolving model, e.g., Grabowski, 2004) could lead the way for the development of more realistic cloud fields and thus improved treatments of aerosol-cloud interactions in large-scale models. How-

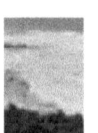

Aerosols can alter the location of solar radiation absorption within the system, and this aspect alone can alter climate and precipitation.

ever, these are just being incorporated in models that resolve both cloud and aerosols. Detailed cloud parcel models have been developed to focus on the droplet activation problem (that asks under what conditions droplets actually start forming) and questions associated with the first indirect effect. The coupling of aerosol and cloud modules to dynamical models that resolve the large turbulent eddies associated with vertical motion and clouds [large eddy simulations (LES) models, with grid sizes of ~100 m and domains ~10 km] has proven to be a powerful tool for representing the details of aerosol-cloud interactions together with feedbacks (e.g., Feingold et al., 1994; Kogan et al., 1994; Stevens et al., 1996; Feingold et al., 1999; Ackerman et al., 2004). This section explores some of the complexity in the aerosol indirect effects revealed by such studies to illustrate how difficult parameterizing these effects properly in GCMs could really be.

3.4.4A. The First Indirect Effect

The relationship between aerosol and drop concentrations (or drop sizes) is a key piece of the first indirect effect puzzle. (It should not, however, be equated to the first indirect effect which concerns itself with the resultant RF). A huge body of measurement and modeling work points to the fact that drop concentrations increase with increasing aerosol. The main unresolved questions relate to the degree of this effect, and the relative importance of aerosol size distribution, composition and updraft velocity in determining drop concentrations (for a review, see McFiggans et al., 2006). Studies indicate that the aerosol number concentration and size distribution are the most important aerosol factors. Updraft velocity (unresolved by GCMs) is particularly important under conditions of high aerosol particle number concentration.

Although it is likely that composition has some effect on drop number concentrations, composition is generally regarded as relatively unimportant compared to the other parameters (Fitzgerald, 1975; Feingold, 2003; Ervens et al., 2005; Dusek et al., 2006). Therefore, it has been stated that the significant complexity in aerosol composition can be modeled, for the most part, using fairly simple parameterizations that reflect the soluble and insoluble fractions (e.g., Rissler et al., 2004). However, composition cannot be simply dismissed. Furthermore, chemical interactions also cannot be overlooked. A large

uncertainty remains concerning the impact of organic species on cloud droplet growth kinetics, thus cloud droplet formation. Cloud drop size is affected by wet scavenging, which depends on aerosol composition especially for freshly emitted aerosol. And future changes in composition will presumably arise due to biofuels/biomass burning and a reduction in sulfate emissions, which emphasizes the need to include composition changes in models when assessing the first indirect effect. The simple soluble/insoluble fraction model may become less applicable than is currently the case.

The updraft velocity, and its change as climate warms, may be the most difficult aspect to simulate in GCMs because of the small scales involved. In GCMs it is calculated in the dynamics as a grid box average, and parameterized on the small scale indirectly because it is a key part of convection and the spatial distribution of condensate, as well as droplet activation. Numerous solutions to this problem have been sought, including estimation of vertical velocity based on predicted turbulent kinetic energy from boundary layer models (Lohmann et al., 1999; Larson et al., 2001) and PDF representations of subgrid quantities, such as vertical velocity and the vertically-integrated cloud liquid water ('liquid water path', or LWP) (Pincus and Klein, 2000; Golaz et al., 2002a,b; Larson et al., 2005). Embedding cloud-resolving models within GCMs is also being actively pursued (Grabowski et al., 1999; Randall et al., 2003). Numerous other details come into play; for example, the treatment of cloud droplet activation in GCM frameworks is often based on the assumption of adiabatic conditions, which may overestimate the sensitivity of cloud to changes in CCN (Sotiropoulou et al., 2006, 2007). This points to the need for improved theoretical understanding followed by new parameterizations.

3.4.4B. Other Indirect Effects

The second indirect effect is often referred to as the "cloud lifetime effect", based on the premise that non-precipitating clouds will live longer. In GCMs the "lifetime effect" is equivalent to changing the representation of precipitation production and can be parameterized as an increase in cloud area or cloud cover (e.g., Hansen et al., 2005). The second indirect effect hypothesis states that the more numerous and smaller drops associated with aerosol perturbations, suppress

collision-induced rain, and result in a longer cloud lifetime. Observational evidence for the suppression of rain in warm clouds exists in the form of isolated studies (e.g. Warner, 1968) but to date there is no statistically robust proof of surface rain suppression (Levin and Cotton, 2008). Results from ship-track studies show that cloud water may increase or decrease in the tracks (Coakley and Walsh, 2002) and satellite studies suggest similar results for warm boundary layer clouds (Han et al., 2002). Ackerman et al. (2004) used LES to show that in stratocumulus, cloud water may increase or decrease in response to increasing aerosol depending on the relative humidity of the air overlaying the cloud. Wang et al. (2003) showed that all else being equal, polluted stratocumulus clouds tend to have lower water contents than clean clouds because the small droplets associated with polluted clouds evaporate more readily and induce an evaporation-entrainment feedback that dilutes the cloud. This result was confirmed by Xue and Feingold (2006) and Jiang and Feingold (2006) for shallow cumulus, where pollution particles were shown to decrease cloud fraction. Furthermore, Xue et al. (2008) suggested that there may exist two regimes: the first, a precipitating regime at low aerosol concentrations where an increase in aerosol will suppress precipitation and increase cloud cover (Albrecht, 1989); and a second, non precipitating regime where the enhanced evaporation associated with smaller drops will decrease cloud water and cloud fraction.

The possibility of bistable aerosol states was proposed earlier by Baker and Charlson (1990) based on consideration of aerosol sources and sinks. They used a simple numerical model to suggest that the marine boundary layer prefers two aerosol states: a clean, oceanic regime characterized by a weak aerosol source and less reflective clouds; and a polluted, continental regime characterized by more reflective clouds. On the other hand, study by Ackerman et al. (1994) did not support such a bistable system using a somewhat more sophisticated model. Further observations are needed to clarify the nature of cloud/aerosol interactions under a variety of conditions.

Finally, the question of possible effects of aerosol on cloud lifetime was examined by Jiang et al. (2006), who tracked hundreds of cumulus clouds generated by LES from their formative stages until they dissipated. They showed that in the model there was no effect of aerosol on cloud lifetime, and that cloud lifetime was dominated by dynamical variability.

It could be argued that the representation of these complex feedbacks in GCMs is not warranted until a better understanding of the processes is at hand. Moreover, until GCMs are able to represent cloud scales, it is questionable what can be obtained by adding microphysical complexity to poorly resolved clouds. A better representation of aerosol-cloud interactions in GCMs therefore depends on the ability to improve representation of aerosols and clouds, as well as their interaction, in the hydrologic cycle. This issue is discussed further in the next chapter.

3.5. Aerosol in the Climate Models

3.5.1. AEROSOL IN THE IPCC AR4 CLIMATE MODEL SIMULATIONS

To assess the atmospheric and climate response to aerosol forcing, e.g., changes in surface temperate, precipitation, or atmospheric circulation, aerosols, together with greenhouse gases should be an integrated part of climate model simulation under the past, present, and future conditions. Table 3.6 lists the forcing species that were included in 25 climate modeling groups used in the IPCC AR4 (2007) assessment. All the models included long-lived greenhouse gases, most models included sulfate direct forcing, but only a fraction of those climate models considered other aerosol types. In other words, aerosol RF was not adequately accounted for in the climate simulations for the IPCC AR4. Put still differently, the current aerosol modeling capability has not been fully incorporated into the climate model simulations. As pointed out in Section 3.4, fewer than one-third of the models incorporated an aerosol indirect effect, and most considered only sulfates.

The following discussion compares two of the IPCC AR4 climate models that include all major forcing agencies in their climate simulation: the model from the NASA Goddard Institute for Space Studies (GISS) and from the NOAA Geophysical Fluid Dynamics Laboratory (GFDL). The purpose in presenting these comparisons is to help elucidate how modelers go about assessing their aerosol components,

A better representation of aerosol-cloud interactions in climate models depends on the ability to improve representation of aerosols and clouds, as well as their interaction, in the hydrologic cycle.

Table 3.6. Forcings used in IPCC AR4 simulations of 20th century climate change. This table is adapted from SAP 1.1 Table 5.2 (compiled using information provided by the participating modeling centers, see http://www-pcmdi. llnl.gov/ipcc/model_documentation/ipcc_model_documentation.php) plus additional information from that website. Eleven different forcings are listed: well-mixed greenhouse gases (G), tropospheric and stratospheric ozone (O), sulfate aerosol direct (SD) and indirect effects (S), black carbon (BC) and organic carbon aerosols (OC), mineral dust (MD), sea salt (SS), land use/land cover (LU), solar irradiance (SO), and volcanic aerosols (V). Check mark denotes inclusion of a specific forcing. As used here, "inclusion" means specification of a time-varying forcing, with changes on interannual and longer timescales.

	MODEL	COUNTRY	G	O	SD	SI	BC	OC	MD	SS	LU	SO	V
1	BCC-CM1	China	√	√	√								
2	BCCR-BCM2.0	Norway	√		√				√	√			
3	CCSM3	USA	√	√	√		√	√				√	√
4	CGCM3.1(T47)	Canada	√		√								
5	CGCM3.1(T63)	Canada	√		√								
6	CNRM-CM3	France	√	√	√		√						
7	CSIRO-Mk3.0	Australia	√		√								
8	CSIRO-Mk3.5	Australia	√		√								
9	ECHAM5/MPI-OM	Germany	√	√	√	√							
10	ECHO-G	Germany/Korea	√	√	√	√						√	√
11	FGOALS-g1.0	China	√		√								
12	GFDL-CM2.0	USA	√	√	√		√	√			√	√	√
13	GFDL-CM2.1	USA	√	√	√		√	√			√	√	√
14	GISS-AOM	USA	√		√					√			
15	GISS-EH	USA	√	√	√	√	√	√	√	√	√	√	√
16	GISS-ER	USA	√	√	√	√	√	√	√	√	√	√	√
17	INGV-SXG	Italy	√	√	√								
18	INM-CM3.0	Russia	√		√							√	
19	IPSL-CM4	France	√		√	√							
20	MIROC3.2(hires)	Japan	√	√	√		√	√	√	√	√	√	√
21	MIROC3.2(medres)	Japan	√	√	√		√	√	√	√	√	√	√
22	MRI-CGCM2.3.2	Japan	√		√							√	√
23	PCM	USA	√	√	√							√	√
24	UKMO-HadCM3	UK	√	√	√	√							
25	UKMO-HadGEM1	UK	√	√	√	√	√	√			√	√	√

and the difficulties that entail. A particular concern is how aerosol forcings were obtained in the climate model experiments for IPCC AR4. Comparisons with observations have already led to some improvements that can be implemented in climate models for subsequent climate change experiments (e.g., Koch et al., 2006, for GISS model). This aspect is discussed further in chapter 4.

3.5.1A. THE GISS MODEL

There have been many different configurations of aerosol simulations in the GISS model over the years, with different emissions, physics packages, etc., as is apparent from the multiple GISS entries in the preceding figures and tables. There were also three different GISS GCM submissions to IPCC AR4, which varied in their model physics and ocean formulation.

(Note that the aerosols in these three GISS versions are different from those in the AeroCom simulations described in section 3.2 and 3.3.) The GCM results discussed below all relate to the simulations known as GISS model ER (Schmidt et al., 2006, see Table 3.6).

Although the detailed description and model evaluation have been presented in Liu et al. (2006), below are the general characteristics of aerosols in the GISS ER:

Aerosol fields: The aerosol fields used in the GISS ER is a prescribed "climatology" which is obtained from chemistry transport model simulations with monthly averaged mass concentrations representing conditions up to 1990. Aerosol species included are sulfate, nitrate, BC, POM, dust, and sea salt. Dry size effective radii are specified for each of the aerosol types, and laboratory-measured phase functions are employed for all solar and thermal wavelengths. For hygroscopic aerosols (sulfate, nitrate, POM, and sea salt), formulas are used for the particle growth of each aerosol as a function of relative humidity, including the change in density and optical parameters. With these specifications, the AOD, single scattering albedo, and phase function of the various aerosols are calculated. While the aerosol distribution is prescribed as monthly mean values, the relative humidity component of the extinction is updated each hour. The global averaged AOD at 550 nm is about 0.15.

Global distribution: When comparing with AOD from observations by multiple satellite sensors of MODIS, MISR, POLDER, and AVHRR and surface based sunphotometer network AERONET (see Chapter 2 for detailed information about data), qualitative agreement is apparent, with generally higher burdens in Northern Hemisphere summer, and seasonal variations of smoke over southern Africa and South America, as well as wind blown dust over northern African and the Persian Gulf. Aerosol optical depth in both model and observations is smaller away from land. There are, however, considerable discrepancies between the model and observations. Overall, the GISS GCM has reduced aerosol optical depths compared with the satellite data (a global, clear-sky average of about 80% compared with MODIS and MISR data), although it is in better agreement with

AERONET ground-based measurements in some locations (note that the input aerosol values were calibrated with AERONET data). The model values over the Sahel in Northern Hemisphere winter and the Amazon in Southern Hemisphere winter are excessive, indicative of errors in the biomass burning distributions, at least partially associated with an older biomass burning source used (the source used here was from Liousse et al., 1996).

Seasonal variation: A comparison of the seasonal distribution of the global AOD between the GISS model and satellite data indicates that the model seasonal variation is in qualitative agreement with observations for many of the locations that represent major aerosol regimes, although there are noticeable differences. For example, in some locations the seasonal variations are different from or even opposite to the observations.

Particle size parameter: The Ångström exponent ($Å$), which is determined by the contrast between the AOD at two or more different wavelengths and is related to aerosol particle size (discussed in section 3.3). This parameter is important because the particle size distribution affects the efficiency of scattering of both short and long wave radiation, as discussed earlier. $Å$ from the GISS model is biased low compared with AERONET, MODIS, and POLDER data, although there are technical differences in determining the $Å$. This low bias suggests that the aerosol particle size in the GISS model is probably too large. The average effective radius in the GISS model appears to be 0.3-0.4 µm, whereas the observational data indicates a value more in the range of 0.2-0.3 µm (Liu et al., 2006).

Single scattering albedo: The model-calculated SSA (at 550 nm) appears to be generally higher than the AERONET data at worldwide locations (not enough absorption), but lower than AERONET data in Northern Africa, the Persian Gulf, and the Amazon (too much absorption). This discrepancy reflects the difficulties in modeling BC, which is the dominant absorbing aerosol, and aerosol sizes. Global averaged SSA at 550 nm from the GISS model is at about 0.95.

Aerosol direct RF: The GISS model calculated anthropogenic aerosol direct shortwave RF is -0.56 W m⁻² at TOA and -2.87 W m⁻² at the surface. The TOA forcing (upper left, Figure 3.7) indicates that, as expected, the model has larger negative values in polluted regions and positive forcing at the highest latitudes. At the surface (lower left, Figure 3.7) GISS model values exceed -4 W m⁻² over large regions. Note there is also a longwave RF of aerosols (right column), although they are much weaker than the shortwave RF.

There are several concerns for climate change simulations related to the aerosol trend in the GISS model. One is that the aerosol fields in the GISS AR4 climate simulation (version ER) are kept fixed after 1990. In fact, the observed trend shows a reduction in tropospheric aerosol optical thickness from 1990 through the present, at least over the oceans (Mishchenko and Geogdzhayev, 2007). Hansen et al. (2007) suggested that the deficient warming in the GISS model over Eurasia post-1990 was due to the lack of this trend. Indeed, a possible conclusion from the Penner et al. (2002) study was that the GISS model overestimated the AOD (presumably associated with anthropogenic aerosols) poleward of 30°N. However, when an alternate experiment reduced the aerosol optical depths, the polar warming became excessive (Hansen et al., 2007). The other concern is that the GISS model may underestimate the organic and sea salt AOD, and overestimate the influence of black carbon aerosols in the biomass burning regions (deduced from Penner et al., 2002; Liu

et al., 2006). To the extent that is true, it would indicate the GISS model underestimates the aerosol direct cooling effect in a substantial portion of the tropics, outside of biomass burning areas. Clarifying those issues requires numerous modeling experiments and various types of observations.

3.5.1B. THE GFDL MODEL

A comprehensive description and evaluation of the GFDL aerosol simulation are given in Ginoux et al. (2006). Below are the general characteristics:

Aerosol fields: The aerosols used in the GFDL climate experiments are obtained from simulations performed with the MOZART 2 model (Model for Ozone and Related chemical Tracers) (Horowitz et al., 2003; Horozwitz, 2006). The exceptions were dust, which was generated with a separate simulation of MOZART 2, using sources from Ginoux et al. (2001) and wind fields from NCEP/NCAR reanalysis data; and sea salt, whose monthly mean concentrations were obtained from a previous study by Haywood et al. (1999). It includes most of the same aerosol species as in the GISS model (although it does not include nitrates), and, as in the GISS model, relates the dry aerosol to wet aerosol optical depth via the model's relative humidity for sulfate (but not for organic carbon); for sea salt, a constant relative humidity of 80% was used. Although the parameterizations come from different sources, both models maintain a

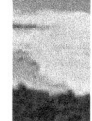

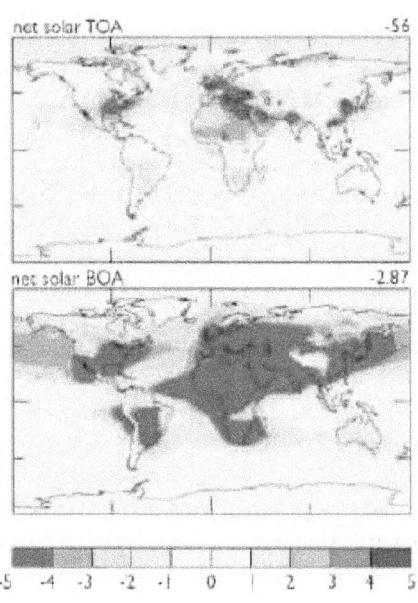

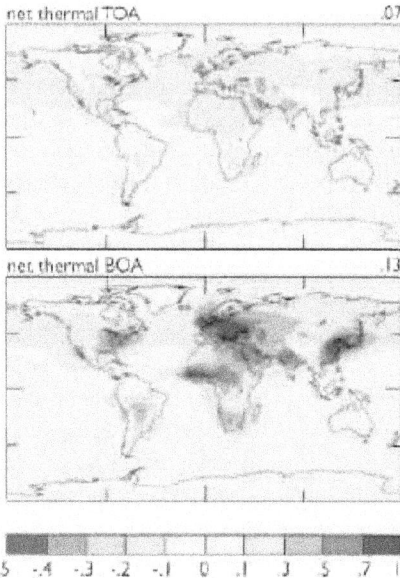

Fig. 3.7. Direct radiative forcing by anthropogenic aerosols in the GISS model (including sulfates, BC, OC and nitrates). Short wave forcing at TOA and surface are shown in the top left and bottom left panels. The corresponding thermal forcing is indicated in the right hand panels. Figure provided by A. Lacis, GISS.

very large growth in sulfate particle size when the relative humidity exceeds 90%.

Global distributions: Overall, the GFDL global mean aerosol mass loading is within 30% of that of other studies (Chin et al., 2002; Tie et al., 2005; Reddy et al., 2005a), except for sea salt, which is 2 to 5 times smaller. However, the sulfate AOD (0.1) is 2.5 times that of other studies, whereas the organic carbon value is considerably smaller (on the order of 1/2). Both of these differences are influenced by the relationship with relative humidity. In the GFDL model, sulfate is allowed to grow up to 100% relative humidity, but organic carbon does not increase in size as relative humidity increases. Comparison of AOD with AVHRR and MODIS data for the time period 1996-2000 shows that the global mean value over the ocean (0.15) agrees with AVHRR data (0.14) but there are significant differences regionally, with the model overestimating the value in the northern mid latitude oceans and underestimating it in the southern ocean. Comparison with MODIS also shows good agreement globally (0.15), but in this case indicates large disagreements over land, with the model producing excessive AOD over industrialized countries and underestimating the effect over biomass burning regions. Overall, the global averaged AOD at 550 nm is 0.17, which is higher than the maximum values in the AeroCom-A experiments (Table 3.2) and exceeds the observed value too (Ae and S* in Figure 3.1).

Composition: Comparison of GFDL modeled species with *in situ* data over North America, Europe, and over oceans has revealed that the sulfate is overestimated in spring and summer and underestimated in winter in many regions, including Europe and North America. Organic and black carbon aerosols are also overestimated in polluted regions by a factor of two, whereas organic carbon aerosols are elsewhere underestimated by factors of 2 to 3. Dust concentrations at the surface agree with observations to within a factor of 2 in most places where significant dust exists, although over the southwest U.S. it is a factor of 10 too large. Surface concentrations of sea salt are underestimated by more than a factor of 2. Over the oceans, the excessive sulfate AOD compensates for the low sea salt values except in the southern oceans.

Size and single-scattering albedo: No specific comparison was given for particle size or single-scattering albedo, but the excessive sulfate would likely produce too high a value of reflectivity relative to absorption except in some polluted regions where black carbon (an absorbing aerosol) is also overestimated.

As in the case of the GISS model, there are several concerns with the GFDL model. The good global-average agreement masks an excessive aerosol loading over the Northern Hemisphere (in particular, over the northeast U.S. and Europe) and an underestimate over biomass burning regions and the southern oceans. Several model improvements are needed, including better parameterization of hygroscopic growth at high relative humidity for sulfate and organic carbon; better sea salt simulations; correcting an error in extinction coefficients; and improved biomass burning emissions inventory (Ginoux et al., 2006).

3.5.1c. Comparisons between GISS and GFDL model

Both GISS and GFDL models were used in the IPCC AR4 climate simulations for climate sensitivity that included aerosol forcing. It would be constructive, therefore, to compare the similarities and differences of aerosols in these two models and to understand what their impacts are in climate change simulations. Figure 3.8 shows the percentage AOD from different aerosol components in the two models.

Sulfate: The sulfate AOD from the GISS model is within the range of that from all other models (Table 3.3), but that from the GFDL model exceeds the maximum value by a factor of 2.5. An assessment in SAP 3.2 (CCSP 2008; Shindell et al., 2008b) also concludes that GFDL had excessive sulfate AOD compared with other models. The sulfate AOD from GFDL is nearly a factor of 4 large than that from GISS, although the sulfate burden differs only by about 50% between the two models. Clearly, this implies a large difference in sulfate MEE between the two models.

BC and POM: Compared to observations, the GISS model appears to overestimate the influence of BC and POM in the biomass burning regions and underestimate it elsewhere, whereas the GFDL model is somewhat the reverse: it overestimates it in polluted regions, and un-

derestimates it in biomass burning areas. The global comparison shown in Table 3.4 indicates the GISS model has values similar to those from other models, which might be the result of such compensating errors. The GISS and GFDL models have relatively similar global-average black carbon contributions, and the same appears true for POM.

Sea salt: The GISS model has a much larger sea salt contribution than does GFDL (or indeed other models).

Global and regional distributions: Overall, the global averaged AOD is 0.15 from the GISS model and 0.17 from GFDL. However, as shown in Figure 3.8, the contribution to this AOD from different aerosol components shows greater disparity. For example, over the Southern Ocean where the primary influence is due to sea salt in the GISS model, but in the GFDL it is sulfate. The lack of satellite observations of the component contributions and the limited available *in situ* measurements make the model improvements at aerosol composition level difficult.

Climate simulations: With such large differences in aerosol composition and distribution between the GISS and GFDL models, one might expect that the model simulated surface temperature might be quite different. Indeed, the GFDL model was able to reproduce the observed temperature change during the 20th century without the use of an indirect aerosol effect, whereas the GISS model required a substantial indirect aerosol contribution (more than half of the total aerosol forcing; Hansen et al., 2007). It is likely that the reason for this difference was the excessive direct effect in the GFDL model caused by its overestimation of the sulfate optical depth. The GISS model direct aerosol effect (see Section 3.6) is close to that derived from observations (Chapter 2); this suggests that for models with climate sensitivity close to $0.75°C/(W m^{-2})$ (as in the GISS and GFDL models), an indirect effect is needed.

3.5.2. Additional Considerations

Long wave aerosol forcing: So far only the aerosol RF in the shortwave (solar) spectrum has been discussed. Figure 3.7 (right column) shows that compared to the shortwave forcing, the values of anthropogenic aerosol long wave (thermal) forcing in the GISS model are on the

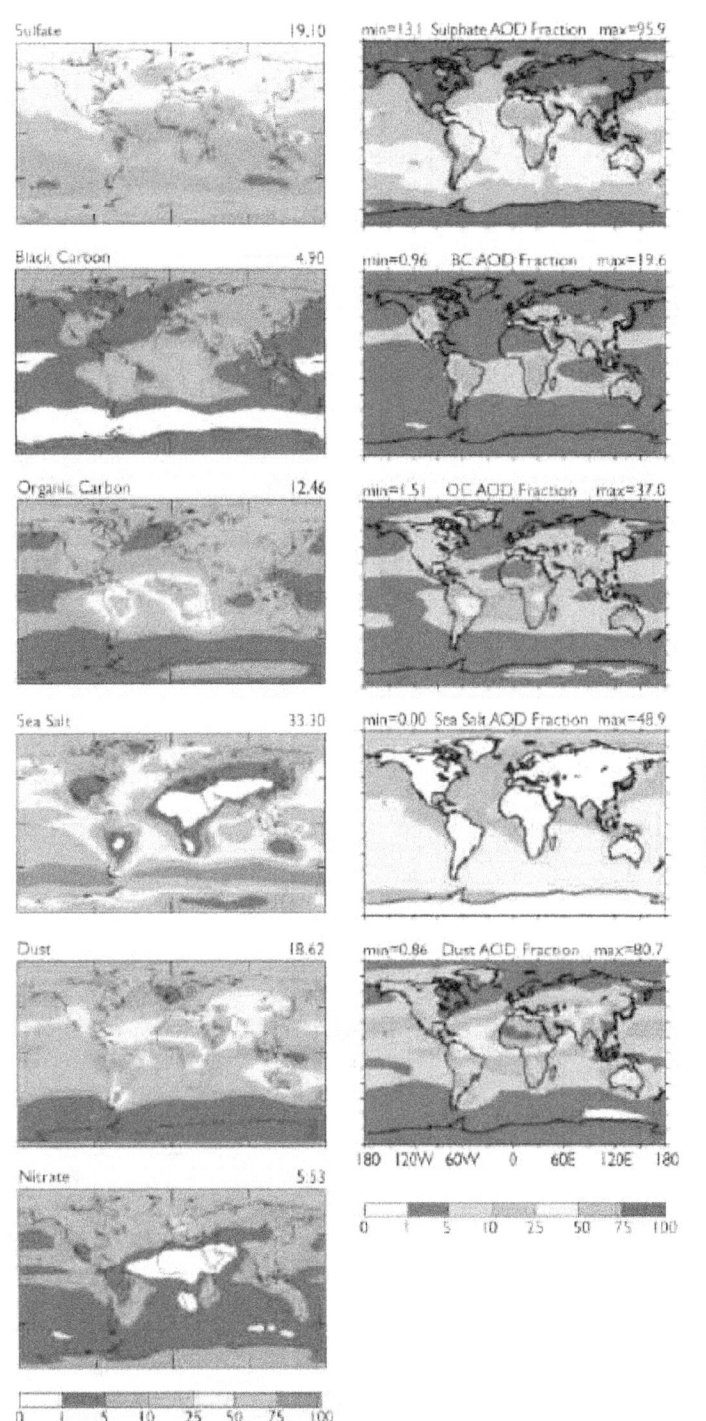

Fig. 3.8. Percentage of aerosol optical depth in the GISS (left, based on Liu et al., 2006, provided by A. Lacis, GISS) and GFDL (right, from Ginoux et al., 2006) models associated with the different components: Sulfate (1st row), BC (2nd row), OC (3rd row), sea-salt (4th row), dust (5th row), and nitrate (last row. Nitrate not available in GFDL model). Numbers on the GISS panels are global average, but on the GFDL panels are maximum and minimum.

order of 10%. Like the shortwave forcing, these values will also be affected by the particular aerosol characteristics used in the simulation.

Aerosol vertical distribution: Vertical distribution is particularly important for absorbing aerosols, such as BC and dust in calculating the RF, particularly when longwave forcing is considered (e.g. Figure 3.7) because the energy they reradiate depends on the temperature (and hence altitude), which affects the calculated forcing values. Several model inter-comparison studies have shown that the largest difference among model simulated aerosol distributions is the vertical profile (e.g. Lohmann et al., 2001; Penner et al., 2002; Textor et al., 2006), due to the significant diversities in atmospheric processes in the models (e.g., Table 3.2). In addition, the vertical distribution also varies with space and time, as illustrated in Figure 3.9 from the GISS ER simulations for January and July showing the most probable altitude of aerosol vertical locations. In general, aerosols in the northern hemisphere are located at lower altitudes in January than in July, and vice versa for the southern hemisphere.

Mixing state: Most climate model simulations incorporating different aerosol types have been made using external mixtures, i.e., the evaluation of the aerosols and their radiative properties are calculated separately for each aerosol type (assuming no mixing between different components within individual particles). Observations indicate that aerosols commonly consist of internally mixed particles, and these "internal

mixtures" can have very different radiative impacts. For example, the GISS-1 (internal mixture) and GISS-2 (external mixture) model results shows very different magnitude and sign of aerosol forcing from slightly positive (implying slight warming) to strong negative (implying significant cooling) TOA forcing (Figure 3.2), due to changes in both radiative properties of the mixtures, and in aerosol amount. The more sophisticated aerosol mixtures from detailed microphysics calculations now being used/developed by different modeling groups may well end up producing very different direct (and indirect) forcing values.

Cloudy sky vs. clear sky: The satellite or AERONET observations are all for clear sky only because aerosol cannot be measured in the remote sensing technique when clouds are present. However, almost all the model results are for all-sky because of difficulty in extracting cloud-free scenes from the GCMs. So the AOD comparisons discussed earlier are not completely consistent. Because AOD can be significantly amplified when relative humidity is high, such as near or inside clouds, all-sky AOD values are expected to be higher than clear sky AOD values. On the other hand, the aerosol RF at TOA is significantly lower for all-sky than for clear sky conditions; the IPCC AR4 and AeroCom RF study (Schulz et al., 2006) have shown that on average the aerosol RF value for all-sky is about 1/3 of that for clear sky although with large diversity (63%). These aspects illustrate the complexity of the system and the difficulty of representing aerosol radiative influences in climate models whose cloud and aerosol distributions are somewhat problematic. And of course aerosols in cloudy regions can affect the clouds themselves, as discussed in Section 3.4.

3.6. Impacts of Aerosols on Climate Model Simulations
3.6.1. Surface Temperature Change

It was noted in the introduction that aerosol cooling is essential in order for models to produce the observed global temperature rise over the last century, at least models with climate sensitivities in the range of 3°C for doubled CO_2 (or ~0.75°C/(W m^{-2})). The implications of this are discussed here in somewhat more detail.

Hansen et al. (2007) show that in the GISS model, well-mixed greenhouse gases produce

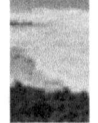

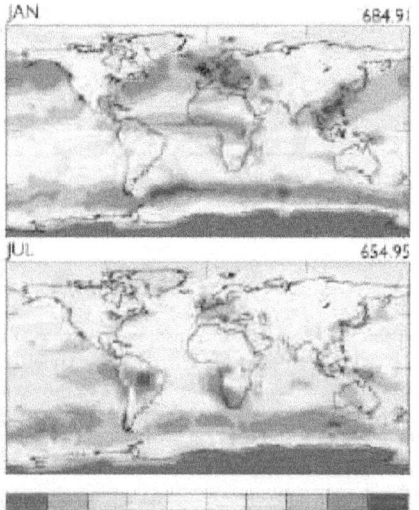

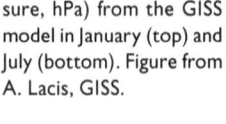

Fig. 3.9. Most probable aerosol altitude (in pressure, hPa) from the GISS model in January (top) and July (bottom). Figure from A. Lacis, GISS.

Table 3.7. Climate forcings (1880-2003) used to drive GISS climate simulations, along with the surface air temperature changes obtained for several periods. Instantaneous (Fi), adjusted (Fa), fixed SST (Fs) and effective (Fe) forcings are defined in Hansen et al., 2005. From Hansen et al., 2007.

Forcing agent	Forcing W m⁻² (1880 – 2003)				ΔT surface °C (year to 2003)			
	Fi	Fa	Fs	Fe	1880	1900	1950	1979
Well-mixed GHGs	2.62	2.50	2.65	2.72	0.96	0.93	0.74	0.43
Stratospheric H_2O			0.06	0.05	0.03	0.01	0.05	0.00
Ozone	0.44	0.28	0.26	0.23	0.08	0.05	0.00	-0.01
Land Use			-0.09	-0.09	-0.05	-0.07	-0.04	-0.02
Snow albedo	0.05	0.05	0.14	0.14	0.03	0.00	0.02	-0.01
Solar Irradiance	0.23	0.24	0.23	0.22	0.07	0.07	0.01	0.02
Stratospheric aerosols	0.00	0.00	0.00	0.00	-0.08	-0.03	-0.06	0.04
Trop. aerosol direct forcing	-0.41	-0.38	-0.52	-0.60	-0.28	-0.23	-0.18	-0.10
Trop. aerosol indirect forcing			-0.87	-0.77	-0.27	-0.29	-0.14	-0.05
Sum of above			**1.86**	**1.90**	**0.49**	**0.44**	**0.40**	**0.30**
All forcings at once			**1.77**	**1.75**	**0.53**	**0.61**	**0.44**	**0.29**

a warming of close to 1°C between 1880 and the present (Table 3.7). The direct effect of tropospheric aerosols as calculated in that model produces cooling of close to -0.3°C between those same years, while the indirect effect (represented in that study as cloud cover change) produces an additional cooling of similar magnitude (note that the general model result quoted in IPCC AR4 is that the indirect RF is twice that of the direct effect).

The time dependence of the total aerosol forcing used as well as the individual species components is shown in Figure 3.10. The resultant warming, 0.53 (±0.04) °C including these and other forcings (Table 3.7), is less than the observed value of 0.6-0.7°C from 1880-2003. Hansen et al. (2007) further show that a reduction in sulfate optical thickness and the direct aerosol effect by 50%, which also reduced the aerosol indirect effect by 18%, produces a negative aerosol forcing from 1880 to 2003 of −0.91 W m⁻² (down from −1.37 W m⁻² with this revised forcing). The model now warms 0.75°C over that time. Hansen et al. (2007) defend this change by noting that sulfate aerosol removal over North America and western Europe during the 1990s led to a cleaner atmosphere. Note that

the comparisons shown in the previous section suggest that the GISS model already underestimates aerosol optical depths; it is thus trends that are the issue here.

The magnitude of the indirect effect used by Hansen et al. (2005) is roughly calibrated to reproduce the observed change in diurnal temperature cycle and is consistent with some satellite observations. However, as Anderson et al. (2003) note, the forward calculation of aerosol negative forcing covers a much larger range than is normally used in GCMs; the values chosen, as in this case, are consistent with the inverse reasoning estimates of what is needed to produce the observed warming, and hence generally consistent with current model climate sensitivities. The authors justify this approach by claiming that paleoclimate data indicate a climate sensitivity of close to 0.75 (±0.25) °C/(W m⁻²), and therefore something close to this magnitude of negative forcing is reasonable. Even this stated range leaves significant uncertainty in climate sensitivity and the magnitude of the aerosol negative forcing. Furthermore, IPCC (2007) concluded that paleoclimate data are not capable of narrowing the range of climate sensitivity, nominally 0.375 to 1.13

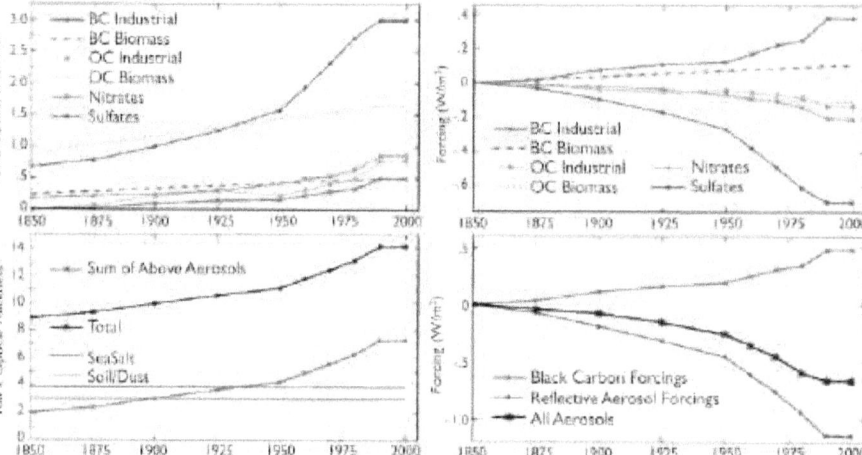

Fig. 3.10. Time dependence of aerosol optical thickness (left) and climate forcing (right). Note that as specified, the aerosol trends are all 'flat' from 1990 to 2000. From Hansen et al. (2007).

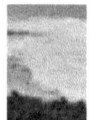

°C/(W m-2), because of uncertainties in paleoclimate forcing and response; so from this perspective the total aerosol forcing is even less constrained than the GISS estimate. Hansen et al. (2007) acknowledge that "an equally good match to observations probably could be obtained from a model with larger sensitivity and smaller net forcing, or a model with smaller sensitivity and larger forcing".

The GFDL model results for global mean ocean temperature change (down to 3 km depth) for the time period 1860 to 2000 is shown in Figure 3.11, along with the different contributing factors (Delworth et al., 2005). This is the same GFDL model whose aerosol distribution was discussed previously. The aerosol forcing produces a cooling on the order of 50% that of greenhouse warming (generally similar to that calculated by the GISS model, Table 3.7). Note that this was achieved without any aerosol indirect effect.

The general model response noted by IPCC, as discussed in the introduction, was that the total aerosol forcing of -1.3 W m-2 reduced the greenhouse forcing of near 3 W m-2 by about 45%, in the neighborhood of the GFDL and GISS forcings. Since the average model sensitivity was close to 0.75 °C/(W m-2), similar to the sensitivities of these models, the necessary negative forcing is therefore similar. The agreement cannot therefore be used to validate the actual aerosol effect until climate sensitivity itself is better known.

Is there some way to distinguish between greenhouse gas and aerosol forcing that would allow the observational record to indicate how much

of each was really occurring? This question of attribution has been the subject of numerous papers, and the full scope of the discussion is beyond the range of this report. It might be briefly noted that Zhang et al. (2006) using results from several climate models and including both spatial and temporal patterns, found that the climate responses to greenhouse gases and sulfate aerosols are correlated, and separation is possible only occasionally, especially at global scales. This conclusion appears to be both model and method-dependent: using time-space distinctions as opposed to trend detection may work differently in different models (Gillett et al., 2002a). Using multiple models helps primarily by providing larger-ensemble sizes for statistics (Gillett et al., 2002b). However, even separating between the effects of different aerosol types is difficult. Jones et al. (2005) concluded that currently the pattern of temperature change due to black carbon is indistinguishable

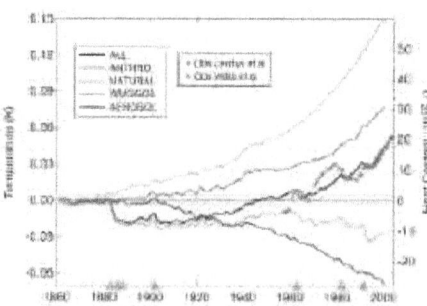

Fig. 3.11. Change in global mean ocean temperature (left axis) and ocean heat content (right axis) for the top 3000 m due to different forcings in the GFDL model. WMGG includes all greenhouse gases and ozone; NATURAL includes solar and volcanic aerosols (events shown as green triangles on the bottom axis). Observed ocean heat content changes are shown as well. From Delworth et al., 2005.

from the sulfate aerosol pattern. In contrast, Hansen et al. (2005) found that absorbing aerosols produce a different global response than other forcings, and so may be distinguishable. Overall, the similarity in response to all these very different forcings is undoubtedly due to the importance of climate feedbacks in amplifying the forcing, whatever its nature.

Distinctions in the climate response do appear to arise in the vertical, where absorbing aerosols produce warming that is exhibited throughout the troposphere and into the stratosphere, whereas reflective aerosols cool the troposphere but warm the stratosphere (Hansen et al., 2005; IPCC, 2007). Delworth et al. (2005) noted that in the ocean, the cooling effect of aerosols extended to greater depths, due to the thermal instability associated with cooling the ocean surface. Hence the temperature response at levels both above and below the surface may provide an additional constraint on the magnitudes of each of these forcings, as may the difference between Northern and Southern Hemisphere changes (IPCC, 2007 Chapter 9). The profile of atmospheric temperature response will be useful to the extent that the vertical profile of aerosol absorption, an important parameter to measure, is known.

3.6.2. Implications for Climate Model Simulations

The comparisons in Sections 3.2 and 3.3 suggest that there are large differences in model calculated aerosol distributions, mainly because of the large uncertainties in modeling the aerosol atmospheric processes in addition to the uncertainties in emissions. The fact that the total optical depth is in better agreement between models than the individual components means that even with similar optical depths, the aerosol direct forcing effect can be quite different, as shown in the AeroCom studies. Because the diversity among models and discrepancy between models and observations are much larger at the regional level than in global average, the assessment of climate response (e.g. surface temperature change) to aerosol forcing would be more accurate for global average than for regional or hemispheric differentiation. However, since aerosol forcing is much more pronounced on regional than on global scales because of the highly variable aerosol distributions, it is insufficient or even misleading to just get the global average right.

The indirect effect is strongly influenced by the aerosol concentrations, size, type, mixing state, microphysical processes, and vertical profile. As shown in previous sections, very large differences exist in those quantities even among the models having similar AOD. Moreover, modeling aerosol indirect forcing presents more challenges than direct forcing because there is so far no rigorous observational data, especially on a global scale, that one can use to test the model simulations. As seen in the comparisons of the GISS and GFDL model climate simulations for IPCC AR4, aerosol indirect forcing was so poorly constrained that it was completely ignored by one model (GFDL) but used by another (GISS) at a magnitude that is more than half of the direct forcing, in order to reproduce the observed surface temperature trends. A majority of the climate models used in IPCC AR4 do not consider indirect effects; the ones that did were mostly limited to highly simplified sulfate indirect effects (Table 3.6). Improvements must be made to at least the degree that the aerosol indirect forcing can no longer be used to mask the deficiencies in estimating the climate response to greenhouse gas and aerosol direct RF.

3.7. Outstanding Issues

Clearly there are still large gaps in assessing the aerosol impacts on climate through modeling. Major outstanding issues and prospects of improving model simulations are discussed below.

Aerosol composition: Many global models are now able to simulate major aerosol types such as sulfate, black carbon, and POM, dust, and sea salt, but only a small fraction of these models simulate nitrate aerosols or consider anthropogenic secondary organic aerosols. And it is difficult to quantify the dust emission from human activities. As a result, the IPCC AR4 estimation of the nitrate and anthropogenic dust TOA forcing was left with very large uncertainty. The next generation of global models should therefore have a more comprehensive suite of aerosol compositions with better-constrained anthropogenic sources.

Aerosol absorption: One of the most critical parameters in aerosol direct RF and aerosol impact on hydrological cycles is the aerosol absorption. Most of the absorption is from BC despite its small contribution to total aerosol

The fact that the total optical depth is in better agreement between models than the individual components means that even with similar optical depths, the aerosol direct forcing effect can be quite different among models... Moreover, modeling aerosol indirect forcing presents more challenges.

load and AOD; dust too absorbs in both the short and long-wave spectral ranges, whereas POM absorbs in the near UV. The aerosol absorption or SSA, will have to be much better represented in the models through improving the estimates of carbonaceous and dust aerosol sources, their atmospheric distributions, and optical properties.

Aerosol indirect effects: The activation of aerosol particles into CCN depends not only on particle size but chemical composition, with the relative importance of size and composition unclear. In current aerosol-climate modeling, aerosol size distribution is generally prescribed and simulations of aerosol composition have large uncertainties. Therefore the model estimated "albedo effect" has large uncertainties. How aerosol would influence cloud lifetime/cover is still in debate. The influence of aerosols on other aspects of the climate system, such as precipitation, is even more uncertain, as are the physical processes involved. Processes that determine aerosol size distributions, hygroscopic growth, mixing state, as well as CCN concentrations, however, are inadequately represented in most of the global models. It will also be difficult to improve the estimate of indirect effects until the models can produce more realistic cloud characteristics.

Aerosol impacts on surface radiation and atmospheric heating: Although these effects are well acknowledged to play roles in modulating atmospheric circulation and water cycle, few coherent or comprehensive modeling studies have focused on them, as compared to the efforts that have gone to assessing aerosol RF at TOA. They have not yet been addressed in the previous IPCC reports. Here, of particular importance is to improve the accuracy of aerosol absorption.

Long-term trends of aerosol: To assess the aerosol effects on climate change the long-term variations of aerosol amount and composition and how they are related to the emission trends in different regions have to be specified. Simulations of historical aerosol trends can be problematic since historical emissions of aerosols have shown large uncertainties—as information is difficult to obtain on past source types, strengths, and even locations. The IPCC AR4 simulations used several alternative aerosol emission histories, especially for BC and POM aerosols.

Climate modeling: Current aerosol simulation capabilities from CTMs have not been fully implemented in most models used in IPCC AR4 climate simulations. Instead, a majority employed simplified approaches to account for aerosol effects, to the extent that aerosol representations in the GCMs, and the resulting forcing estimates, are inadequate. The oversimplification occurs in part because the modeling complexity and computing resource would be significantly increased if the full suite of aerosols were fully coupled in the climate models.

Observational constraints: Model improvement has been hindered by a lack of comprehensive datasets that could provide multiple constraints for the key parameters simulated in the model. The extensive AOD coverage from satellite observations and AERONET measurements has helped a great deal in validating model-simulated AOD over the past decade, but further progress has been slow. Large model diversities in aerosol composition, size, vertical distribution, and mixing state are difficult to constrain, because of lack of reliable measurements with adequate spatial and temporal coverage (see Chapter 2).

Aerosol radiative forcing: Because of the large spatial and temporal differences in aerosol sources, types, emission trends, compositions, and atmospheric concentrations, anthropogenic aerosol RF has profound regional and seasonal variations. So it is an insufficient measure of aerosol RF scientific understanding, however useful, for models (or observation-derived products) to converge only on globally and annually averaged TOA RF values and accuracy. More emphasis should be placed on regional and seasonal comparisons, and on climate effects in addition to direct RF at TOA.

3.8 Conclusions

From forward modeling studies, as discussed in the IPCC (2007), the direct effect of aerosols since pre-industrial times has resulted in a negative RF of about -0.5 ± 0.4 W m^{-2}. The RF due to cloud albedo or brightness effect is estimated to be -0.7 (-1.8 to -0.3) W m^{-2}. Forcing of similar magnitude has been used in some modeling studies for the effect associated with cloud lifetime, in lieu of the cloud brightness influence. The total negative RF due to aerosols according to IPCC (2007) estimates is therefore -1.3 (-2.2 to -0.5) W m^{-2}. With the inverse approach, in which aerosols provide forcing necessary to produce the observed temperature change, values range from -1.7 to -0.4 W m^{-2} (IPCC, 2007). These results represent a substantial advance over previous assessments (e.g., IPCC TAR), as the forward model estimated and inverse approach required aerosol TOA forcing values are converging. However, large uncertainty ranges preclude using the forcing and temperature records to more accurately determine climate sensitivity.

There are now a few dozen models that simulate a comprehensive suite of aerosols. This is done primarily in the CTMs. Model inter-comparison studies have shown that models have merged at matching the global annual averaged AOD observed by satellite instruments, but they differ greatly in the relative amount of individual components, in vertical distributions, and in optical properties. Because of the great spatial and temporal variations of aerosol distributions, regional and seasonal diversities are much larger than that of the global annual mean. Different emissions and differences in atmospheric processes, such as transport, removal, chemistry, and aerosol microphysics, are chiefly responsible for the spread among the models. The varying component contributions then lead to differences in aerosol direct RF, as aerosol scattering and absorption properties depend on aerosol size and type. They also impact the calculated indirect RF, whose variations are further amplified by the wide range of cloud and convective parameterizations in models. Currently, the largest aerosol RF uncertainties are associated with the aerosol indirect effect. Most climate models used for the IPCC AR4 simulations employed simplified approaches, with aerosols specified from stand-alone CTM simulations. Despite the uncertainties in aerosol RF and widely varying model climate sensitivity, the IPCC AR4 models were generally able to reproduce the observed temperature record for the past century. This is because models with lower/higher climate sensitivity generally used less/more negative aerosol forcing to offset the greenhouse gas warming. An equally good match to observed surface temperature change in the past could be obtained from a model with larger climate sensitivity and smaller net forcing, or a model with smaller sensitivity and larger forcing (Hansen et al., 2007). Obviously, both greenhouse gases and aerosol effects have to be much better quantified in future assessments.

Progress in better quantifying aerosol impacts on climate will only be made when the capabilities of both aerosol observations and representation of aerosol processes in models are improved. The primary concerns and issues discussed in this chapter include:

- Better representation of aerosol composition and absorption in the global models
- Improved theoretical understanding of subgrid-scale processes crucial to aerosol-cloud interactions and lifetime
- Improved aerosol microphysics and cloud parameterizations
- Better understanding of aerosol effects on surface radiation and hydrological cycles
- More focused analysis on regional and seasonal variations of aerosols
- More reliable simulations of aerosol historic long-term trends
- More sophisticated climate model simulations with coupled aerosol and cloud processes
- Enhanced satellite observations of aerosol type, SSA, vertical distributions, and aerosol radiative effect at TOA; more coordinated field experiments to provide constraints on aerosol chemical, physical, and optical properties.

A discussion of the "way forward" toward better constraints on aerosol radiative forcing, and hence climate sensitivity, is provided in the next chapter.

Progress in better quantifying aerosol impacts on climate will only be made when the capabilities of both aerosol observations and representation of aerosol processes in models are improved.

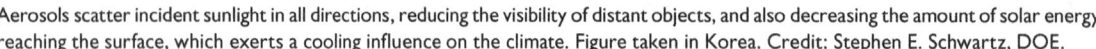

Aerosols scatter incident sunlight in all directions, reducing the visibility of distant objects, and also decreasing the amount of solar energy reaching the surface, which exerts a cooling influence on the climate. Figure taken in Korea. Credit: Stephen E. Schwartz, DOE.

The Way Forward

Authors: David Rind, NASA GISS; Ralph A. Kahn, NASA GSFC; Mian Chin, NASA GSFC; Stephen E. Schwartz, DOE BNL; Lorraine A. Remer, NASA GSFC; Graham Feingold, NOAA ESRL; Hongbin Yu, NASA GSFC/UMBC; Patricia K. Quinn, NOAA PMEL; Rangasayi Halthore, NASA HQ/NRL

4.1. Major Research Needs

This review has emphasized that despite the increase in understanding aerosol forcing of the climate system, many important uncertainties remain. By way of perspective, concerted effort has been directed toward this issue for only about the past 20 years. In view of the variety of aerosol types and emissions, uncertain microphysical properties, great temporal and spatial variability, and the added complexity of aerosol-cloud interactions, it is easy to understand why more work is required to define anthropogenic aerosol forcing with confidence comparable to that for other climate forcing agents.

When comparing surface temperature changes calculated by climate models with those observed, the IPCC AR4 noted "broad consistency" between the modeled and observed temperature record over the industrial period. However, understanding of the degree to which anthropogenic aerosols offset the better-established greenhouse gas forcing is still inadequate. This limits confidence in the predicted magnitude of climate response to future changes in greenhouse gases and aerosols.

This chapter briefly summarizes the major research needs that have been highlighted in previous chapters, recognizing that achieving them will not necessarily be easy or straightforward. Although some important accomplishments will likely be possible in the next decade, others may, realistically, take considerably

longer. Several important points should be kept in mind:

1. The uncertainty in assessing anthropogenic and aerosol impacts on climate must be much reduced from its current level to allow meaningful projections of future climate. Using statistical methods, IPCC AR4 concluded that the present-day global-average anthropogenic RF is 2.9 ± 0.3 W m^{-2} for long-lived greenhouse gases plus ozone, -1.3 (-2.2 to -0.5) W m^{-2} for aerosol direct plus aerosol-cloud-albedo, and +1.6 (0.6 to 2.4) W m^{-2} for total anthropogenic forcing (Figure 1.3 in Chapter 1). As shown in Chapter 1, the current estimate of total anthropogenic RF yields the transient climate sensitivity range of 0.3 – 1.1°C/(W m^{-2}). This translates to a possible surface temperature increase from 1.2°C to 4.7°C at the time of (equivalent) doubled CO_2 forcing, which will likely occur toward the latter part of this century. Such a range is too wide to meaningfully predict the climate response to increased greenhouse gases.

    The large uncertainty in total anthropogenic forcing arises primarily from current uncertainty in the current understanding of aerosol RF, as illustrated in Figure 1.3. One objective should be to reduce the uncertainty in global average RF by anthropogenic aerosols over the industrial period to ±0.3 W m^{-2}, equal to the current uncertainty in

The uncertainty in assessing anthropogenic aerosol impacts on climate must be much reduced from its current level to allow meaningful projections of future climate.

RF by anthropogenic greenhouse gases over this period. Then, taking the total anthropogenic forcing as the IPCC central value, 1.6 W m^{-2}, the range in transient climate sensitivity would be reduced to 0.37 – 0.54°C/(W m^{-2}), and the corresponding increase in global mean surface temperature change at the time of doubled CO_2 forcing would be between 1.5°C and 2.2°C. This range is small enough to make more meaningful global predictions pertinent to planning for mitigation and adaptation.

2. Evaluation of aerosol effects on climate must take into account high spatial and temporal variation of aerosol amounts and properties. Determining the global mean aerosol TOA RF is necessary but far from sufficient, because of the large spatial and temporal variation of aerosol distributions and composition that is in contrast to the much more uniformly distributed longer-lived greenhouse gases such as CO_2 and methane. Therefore, aerosol RF at local to regional scales can be much stronger than its global average.

3. Understanding of the aerosol effects on global water cycle requires much improvement. Besides the radiative forcing, aerosols have other important climate effects. Their heating the atmosphere and cooling the surface affect atmospheric circulations and water cycle. The level of scientific understanding of these effects is much lower than that for aerosol direct RF; concerted research effort is required to move forward.

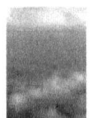

The approach taken for assessing aerosol forcing of the climate system includes both measurement and modeling components. As discussed in Chapters 2 and 3, improved observations, with some assistance from models, are already helping produce measurement-based estimates of the current aerosol direct effect on climate. Global models are now converging on key parameters such as AOD, and thanks to satellite and other atmospheric measurements, are moving toward better assessments of present-day aerosol RF. However, given the relatively short history of satellite observations and the nature of future climate prediction, the assessment of anthropogenic aerosol climate impact for past and future times will inevitably depend on models. Models are also required to apportion observed aerosols between natural and anthropogenic sources. Therefore, improving model predictions of aerosol climate forcing is the key to progress. To do so, it is essential to advance the current measurement capabilities that will allow much better validation of the models and fundamental improvement of model components.

The accuracy of regional to global-scale AOD measured by satellites is currently poorer than needed to substantially reduce uncertainty in direct radiative forcing by aerosols, but the required capability is within reach, based on the accuracy of current local surface-based measurement techniques. Problems remain in converting total aerosol forcing to forcing by anthropogenic aerosols. The accuracy of aerosol vertical distributions as measured by Lidar from space is approaching that required to be useful for evaluating chemical transport models, and is within reach of that required to reduce uncertainties in aerosol direct radiative forcing.

Measurement accuracy for remotely sensed aerosol optical and physical properties (e.g., SSA, g, size) is poorer than needed to significantly reduce uncertainty in aerosol direct radiative forcing and to effect satisfactory translation between AOD retrieved from radiation-based remote-sensing measurements and AOD calculated from CTMs based on aerosol mass concentrations (the fundamental quantities tracked in the model) and optical properties. Combinations of remote-sensing and targeted *in situ* measurement with modeling are required for near-term progress in this area.

Measurements for aerosol indirect effect remain a major challenge. Sensitivity of remote-sensing measurement to particle size, composition, concentration, vertical distribution, and horizontal distribution in the vicinity of clouds is poor. Combinations of detailed *in situ* and laboratory measurements and cloud-resolved modeling, along with spatial extrapolation using remote-sensing measurements and larger-scale modeling, are required for near-term progress in this area.

The next sections address the priorities and recommend approach to moving forward.

4.2. Priorities
4.2.1. MEASUREMENTS

Maintain current and enhance the future satellite aerosol monitoring capabilities. Satellites have been providing global aerosol observations since the late 1970s, with much improved accuracy measurements since late 1990s, but some of them, such as the NASA EOS satellites (Terra, Aqua, Aura), are reaching or exceeding their design lives. Timely follow-on missions to at least maintain these capabilities are important. Assessment of aerosol climate impacts requires a long-term data record having consistent accuracy and high quality, suitable for detecting changes in aerosol amount and type over decadal time scales. Future satellite sensors should have the capability of ac-

quiring information on aerosol size distribution, absorption, vertical distribution, and type with sufficiently high accuracy and adequate spatial coverage and resolution to permit quantification of forcing to required accuracy. The separation of anthropogenic from natural aerosols, perhaps based on size and shape, is essential for assessing human impacts. A brief summary of current capabilities and future needs of major aerosol measurement requirements from space is provided in Table 4.1. (More detailed discussion is in Chapter 2.)

Maintain, enhance, and expand the surface observation networks. Long-term surface-based networks such as the NASA AERONET network, the NOAA ESRL and the DOE ARM sites have for more than a decade been providing essential information on aerosol properties

Table 4.1. Summary of current status and future needs of major aerosol measurements from space for characterization of tropospheric aerosol and determination of aerosol climate forcing.

Satellite instrument	Time Period	AOD	Size or Shape[1]	Absorption[2]	Vertical Profile	Global Coverage
Historic / Current						
AVHRR	Since 1981	√	√			Ocean only
TOMS	1979-2001	√		√		√
POLDER	Since 1997	√	√			√
MODIS	Since 2000	√	√			√
MISR	Since 2000	√	√	√		√
OMI	Since 2004	√		√		√
GLAS	Since 2003[3]		√		√	
CALIOP	Since 2006		√		√	
Scheduled to Launch						
VIIRS (on NPP/NPOESS)	2009-	√				√
OMPS (on NPP)	2009-	√		√		√
APS (on Glory)	2009-	√	√	√		
HSRL (on EarthCARE)	2013-				√	
Future Needs						
Next generation instruments (polarimeter, lidar, etc.) with much improved detection accuracy and coverage for AOD and absorption, enhanced capability for measuring vertical profiles, aerosol types and properties, augmented capacity with measurements of aerosol, clouds, and precipitation.						

[1] Size is inferred from the spectral variation of AOD, expressed as the Ångström exponent.
[2] Determination of absorption from MISR is conditional and not always available.
[3] Aerosol detection by GLAS is limited to only a few months each year because of laser power problems.

that is vital for satellite validation, model evaluation, and climate change assessment from trend analysis. Observation should be enhanced with additional, routine measurements of size-resolved composition, more lidar profiling of vertical features, and improved measurements of aerosol absorption with state-of-art techniques. This, along with climate-quality data records constructed from satellites, would help establish connections between aerosol trends and the observed trends in radiation (e.g., dimming or brightening).

Execute a continuing series of coordinated field campaigns. These would aim to: (1) broaden the database of detailed particle optical, physical, and chemical (including cloud-nucleating) properties for major aerosol types, (2) refine and validate satellite and surface-based remote-sensing retrieval algorithms, (3) make comprehensive, coordinated, multi-platform measurements characterizing aerosols, radiation fields, cloud properties and related aerosol-cloud interactions, to serve as testbeds for modeling experiments at several scales, and (4) deepen the links between the aerosol (and cloud) measuring and modeling communities. New and improved instrument capabilities will be needed to provide more accurate measurements of aerosol absorption and scattering properties across the solar spectrum.

Initiate and carry out a systematic program of simultaneous measurement of aerosol, clouds, and precipitation variables. Measurements of aerosol properties must go hand in hand with measurements of cloud properties, and also with measurements of precipitation and meteorological variables, whether this will be from aircraft, ground-based remote sensing or satellite. Assessing aerosol effects on climate has focused on the interactions of aerosol with Earth's radiation balance (i.e., radiative forcing), but in the near future, focus will shift to include aerosol effects on precipitation patterns, atmospheric circulation, and weather.

Fully exploit the existing information in satellite observations of AOD and particle type. An immense amount of data has been collected. Table 4.1 lists the most widely used aerosol property data sets retrieved from satellite sensors. A synthesis of data from multiple sensors would in many cases be a more effec-

tive resource for aerosol characterizing than data from individual sensors alone. However, techniques for achieving such synthesis are still in their infancy, and multi-sensor products have only begun to be developed. The full information content of existing data, even with individual sensors, has not been realized. There is a need to: (1) refine retrieval algorithms and extract greater information about aerosols from the joint data sets, (2) quantify data quality, (3) generate uniform (and as appropriate, merged), climate-quality data records. These must be applied to: (4) initialize, constrain, and validate models, (5) conduct detailed process studies, and (6) perform statistical trend analysis.

Measure the formation, evolution, and properties of aerosols under controlled laboratory conditions. Laboratory studies are essential to determine chemical transformation rates for aerosol particle formation. They can also provide information, in a controlled environment, for particle hygroscopic growth, light scattering and absorption properties, and particle activation for aerosols of specific, known composition. Such measurements will allow development of suitable mixing rules and evaluation of the parameterizations that rely on such mixing rules.

Improve measurement-based techniques for distinguishing anthropogenic from natural aerosols. Current satellite-based estimates of anthropogenic aerosol fraction rely on retrievals of aerosol type. These estimates suffer from limited information content of the data under many circumstances. More needs to be done to combine satellite aerosol type and vertical distribution retrievals with supporting information from: (1) back-trajectory and inverse modeling, (2) at least qualitative time-series of plume evolution from geosynchronous satellite imaging, and (3) surface monitoring and particularly targeted aircraft *in situ* measurements. Different definitions of "anthropogenic" aerosols will require reconciliation. The anthropogenic fraction of today's aerosol, estimated from current measurements, will not produce the same aerosol radiative forcing defined as the perturbation of the total aerosol from pre-industrial times. Consistently defined perturbation states are required before measurement-based and model-based aerosol radiative forcing estimates can be meaningfully compared.

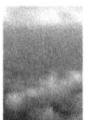

4.2.2. MODELING

Improve the accuracy and capability of model simulation of aerosols (including components and atmospheric processes) and aerosol direct radiative forcing. Spatial and temporal distributions of aerosol mass concentrations are affected primarily by sources, removal mechanisms, atmospheric transport, and chemical transformations; calculations of aerosol direct RF require additional information about aerosol optical properties. Coordinated studies are needed to understand the importance of individual processes, especially vertical mixing and removal by convection/precipitation. Observational strategies must be developed to constrain and validate the key parameters describing: (a) aerosol composition, (b) mass concentration, (c) vertical distribution, (d) size distribution, (e) hygroscopic growth, (f) aerosol absorption, (g) asymmetry parameter and (h) aerosol optical depth. As many models now include major aerosol types such as sulfate, BC, primary POM, dust, and sea salt, progress is needed on simulating nitrate and secondary organic aerosols. In addition, aerosol microphysical processes should be much better represented in the models. In practice, improving the capability of aerosol composition modeling will require improved remote sensing and in situ observations to discriminate among aerosol components. Improvement in modeling radiative forcing could be aided by data assimilation methods, in which the observed aerosol distributions that are input to the model, and the modeled short-term response, could be compared directly with RF observations.

Advance the ability to model aerosol-cloud-precipitation interaction in climate models, particularly the simulation of clouds. The interaction between aerosols and clouds is probably the biggest uncertainty of all climate forcing/feedback processes. The processes involved are complex, and accurate simulation will require sub-grid calculations or improved aerosol and cloud parameterizations on global-model scales. Among the key elements required are: (a) cloud nucleating properties for different aerosol types and size distributions, (b) CCN concentrations as functions of supersaturation and any kinetic influences, (c) algorithms to simulate aerosol influences on cloud brightness, that include cloud fraction, cloud liquid water content, and precipitation efficiency, and (d) cloud drop concentration for known (measured) updraft, humidity, and temperature conditions. Improved aerosol-cloud interaction modeling must be built upon more realistic simulation of clouds and cloud process in GCMs. Cloud-resolving models offer one approach to tackling these questions, aided by the continual improvement in computing capability that makes possible simulations at the higher resolutions appropriate to these processes. Realizing the latter approach, however, may be a long-term goal.

Incorporate improved representation of aerosol processes in coupled aerosol-climate system models to assess the climate change. Coupling aerosol processes in the GCMs would represent a major step in climate simulation beyond the IPCC AR4. This would enable aerosols to interact with the meteorological variables such as clouds and precipitation. Climate change simulations need to be run for hundreds of years with coupled atmosphere-ocean models. Inclusion of aerosol physics and chemistry, and increasing the model resolution, will put large demands on computing power and resources. Some simplification may be necessary, especially considering that other required model improvements, such as finer resolution and carbon cycle models, also increase computing time. The near-term step is to include simple representations of aerosols directly in climate models, incorporating the major aerosol types, basic chemistry, and parameterized cloud droplet activation schemes. Such models exist today, and are ready to be applied to long-term simulations, making it possible to calculate first-order aerosol climate feedbacks. The next generation of models will include aerosol processes that allow for more realistic interactions, such as aerosol and cloud microphysical processes; however, the complexity included should be commensurate with that for other relevant components of the simulation, such as clouds and convection. Fully coupled aerosol-chemistry-physics-climate models will likely be a model-development focus for at least the next decade. This should eventually lead to increasingly sophisticated model simulations of aerosol effects on climate, and better assessments of climate sensitivity.

4.2.3. EMISSIONS

Develop and evaluate emissions inventories of aerosol particles and precursor gases. A systematic determination of emissions of primary particles and of aerosol precursor gases is needed as input to modeling the geographical and temporal distribution of the amount and radiative forcing of aerosols. The required description of emissions includes the location, timing, activity, and amount. For particles, the emissions should be characterized by size distributed composition, not simply by mass emissions, because of the effects these properties have on direct and indirect forcings. Natural emissions from biogenic and volcanic sources should be systematically assessed. Satellite fire data are now being used to help constrain biomass-burning emissions, which include new information on aerosol injection height. Dust emission from human activities, such as from farming practices and land-use changes, likewise needs to be quantified. Characterization of aerosol trends and radiative forcing also requires historical emission data. For assessing anthropogenic impacts on future climate, projections of future anthropogenic fuel use and changes in wildfire, desert dust, biogenic, and other sources are needed, and methods used to obtain them carefully evaluated and possibly refined. Some such efforts are being pursued in conjunction with the IPCC.

4.3. Concluding Remarks

Narrowing the gap between the current understanding of the contribution of anthropogenic aerosols to radiative forcing and that of the long lived greenhouse gases will require progress in all aspects of aerosol-climate science. Development of new space-based, field, and laboratory instruments will be needed, and in parallel, more realistic simulations of aerosol, cloud, and atmospheric processes must be incorporated into models. Most importantly, greater synergy among different types of measurements, different types of models, and especially between measurements and models,

is critical. Aerosol-climate science must expand to encompass not only radiative effects on climate, but also aerosol effects on cloud processes, precipitation, and weather. New initiatives will strive to more effectively include experimentalists, remote sensing scientists and modelers as equal partners, and the traditionally defined communities of aerosol scientists, cloud scientists, radiation scientists increasingly will find common ground in addressing the challenges ahead.

Narrowing the gap between the current understanding of the contribution of anthropogenic aerosols to radiative forcing and that of the long lived greenhouse gases will require progress in all aspects of aerosol-climate science.

Greater synergy among different types of measurements, different types of models, and especially between measurements and models, is critical.

Intense heat generated by the wildfires below causes smoke to rise in violent updrafts; updrafts and water vapor emanating from the fires also combine to produce towering pyrocumulus clouds. Photo taken from the NASA P-3B aircraft during the ARCTAS field experiment in July 2008 over Canada. Credit: Cameron McNaughton, University of Hawaii.

GLOSSARY AND ACRONYMS

GLOSSARY

(Note: Terms in *italic* in each paragraph are defined elsewhere in this glossary.)

Absorption
the process in which incident radiant energy is retained by a substance.

Absorption coefficient
fraction of incident radiant energy removed by *absorption* per length of travel of radiation through the substance.

Active remote sensing
a remote sensing system that transmits its own energy source, then measures the properties of the returned signal. Contrasted with *passive remote sensing*.

Adiabatic equilibrium
a vertical distribution of temperature and pressure in an atmosphere in hydrostatic equilibrium such that an air parcel displaced adiabatically will continue to possess the same temperature and pressure as its surroundings, so that no restoring force acts on a parcel displaced vertically.

Aerosol
a colloidal suspension of liquid or solid particles (in air).

Aerosol asymmetry factor (also called asymmetry parameter, *g*)
the mean cosine of the scattering angle, found by integration over the complete scattering *phase function* of aerosol; $g = 1$ denotes completely forward scattering and $g = 0$ denotes symmetric scattering. For spherical particles, the asymmetry parameter is related to particle size in a systematic way: the larger the particle size, the more the scattering in the forward hemisphere.

Aerosol direct radiative effect
change in radiative flux due to aerosol scattering and absorption with the presence of aerosol relative to the absence of aerosol.

Aerosol hemispheric backscatter fraction (*b*)
the fraction of the scattered intensity that is redirected into the backward hemisphere relative to the incident light; can be determined from measurements made with an integrating nephelometer. The larger the particle size, the smaller the *b*.

Aerosol indirect effects
processes referring to the influence of aerosol on cloud droplet concentration or radiative properties. Effects include the effect of aerosols on cloud droplet size and therefore its brightness (also known as the "cloud albedo effect", "first aerosol indirect effect", or "Twomey effect"); and the effect of cloud droplet size on precipitation efficiency and possibly cloud lifetime (also known as the "second aerosol indirect effect" or "Albrecht effect").

Aerosol mass extinction (scattering, absorption) efficiency
the aerosol *extinction* (*scattering*, *absorption*) *coefficient* per aerosol mass concentration, with a commonly used unit of $m^2\,g^{-1}$.

Aerosol optical depth
the (wavelength dependent) negative logarithm of the fraction of radiation (or light) that is extinguished (or *scattered* or *absorbed*) by *aerosol* particles on a vertical path, typically from the surface (or some specified altitude) to the top of the atmosphere. Alternatively and equivalently: The (dimensionless) line integral of the *absorption coefficient* (due to aerosol particles), or of the *scattering coefficient* (due to aerosol particles), or of the sum of the two (*extinction coefficient* due to aerosol particles), along such a vertical path. Indicative of the amount of aerosol in the column, and specifically relates to the magnitude of interaction between the aerosols and *shortwave* or *longwave radiation*.

Aerosol phase function
the angular distribution of radiation scattered by aerosol particle or by particles comprising an *aerosol*. In practice, the phase function is parameterized with *asymmetry factor* (or *asymmetry parameter*). Aerosol phase function is related to *aerosol hemispheric backscatter fraction* (*b*) and aerosol particle size: the larger the particle size, the more the forward *scattering* (i.e. larger *g* and smaller *b*).

Aerosol radiative forcing
the net energy flux (downwelling minus upwelling) difference between an initial and a perturbed aerosol loading state, at a specified level in the atmosphere. (Other quantities, such as solar radiation, are assumed to be the same.) This difference is defined such that a negative aerosol forcing implies that the change in aerosols relative to the initial state exerts a cooling in-

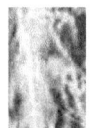

fluence, whereas a positive forcing would mean the change in aerosols exerts a warming influence. The aerosol radiative forcing must be qualified by specifying the initial and perturbed aerosol states for which the radiative flux difference is calculated, the altitude at which the quantity is assessed, the wavelength regime considered, the temporal averaging, the cloud conditions, and whether total or only human-induced contributions are considered (see Chapter 1, Section 1.2).

Aerosol radiative forcing efficiency

aerosol direct radiative forcing per *aerosol optical depth* (usually at 550 nm). It is governed mainly by aerosol size distribution and chemical composition (determining the aerosol *single-scattering albedo* and *phase function*), surface reflectivity, and solar irradiance.

Aerosol semi-direct effect

the processes by which *aerosols* change the local temperature and moisture (e.g., by direct radiative heating and changing the heat releases from surface) and thus the local relative humidity, which leads to changes in cloud liquid water and perhaps cloud cover.

Aerosol single-scattering albedo (SSA)

a ratio of the *scattering coefficient* to the *extinction coefficient* of an aerosol particle or of the particulate matter of an aerosol. More absorbing aerosols and smaller particles have lower SSA.

Aerosol size distribution

probability distribution function of the number concentration, surface area, or volume of the particles comprising an aerosol, per interval (or logarithmic interval) of radius, diameter, or volume.

Albedo

the ratio of reflected flux density to incident flux density, referenced to some surface; might be Earth surface, top of the atmosphere.

Angström exponent ($\mathring{A}$)

exponent that expresses the spectral dependence of *aerosol optical depth* (τ) (or *scattering coefficient*, absorption *coefficient*, etc.) with the wavelength of light (λ) as inverse power law: $\tau \propto \lambda^{-A}$. The Ångström exponent is inversely related to the average size of aerosol particles: the smaller the particles, the larger the exponent.

Anisotropic

not having the same properties in all directions.

Atmospheric boundary layer (abbreviated **ABL**; also called planetary boundary layer—**PBL**)

the bottom layer of the troposphere that is in contact with the surface of the earth. It is often turbulent and is capped by a statically stable layer of air or temperature inversion. The ABL depth (i.e., the inversion height) is variable in time and space, ranging from tens of meters in strongly statically stable situations, to several kilometers in convective conditions over deserts.

Bidirectional reflectance distribution function (BRDF)

a relationship describing the reflected radiance from a given region as a function of both incident and viewing directions. It is equal to the reflected *radiance* divided by the incident *irradiance* from a single direction.

Clear-sky radiative forcing

radiative forcing (of gases or aerosols) in the absence of clouds. Distinguished from total-sky or all-sky *radiative forcing*, which include both cloud-free and cloudy regions.

Climate sensitivity

the change in global mean near-surface temperature per unit of *radiative forcing*; when unqualified typically refers to equilibrium sensitivity; transient sensitivity denotes time dependent change in response to a specified temporal profile.

Cloud albedo

the fraction of solar radiation incident at the top of cloud that is reflected by clouds in the atmosphere or some subset of the atmosphere.

Cloud condensation nuclei (abbreviated **CCN**)

aerosol particles that can serve as seed particles of atmospheric cloud droplets, that is, particles on which water condenses (activates) at *supersaturations* typical of atmospheric cloud formation (fraction of one percent to a few percent, depending on cloud type); may be specified as function of supersaturation.

Cloud resolving model

a numerical model that resolves cloud-scale (and mesoscale) circulations in three (or sometimes two) spatial dimensions. Usually run with horizontal resolution of 5 km or less.

Coalescence

the merging of two or more droplets of precipitation (or aerosol particles; also denoted coagulation) into a single droplet or particle.

Condensation

in general, the physical process (phase transition) by which a vapor becomes a liquid or solid; the opposite of *evaporation*.

Condensation nucleus (abbreviated **CN**)

an aerosol particle forming a center for *condensation* under extremely high *supersaturations* (up to 400% for water, but below that required to activate small ions).

Data assimilation
the combining of diverse data, possibly sampled at different times and intervals and different locations, into a unified and physically consistent description of a physical system, such as the state of the atmosphere.

Diffuse radiation
radiation that comes from some continuous range of directions. This includes radiation that has been scattered at least once, and emission from nonpoint sources.

Dry deposition
the process by which atmospheric gases and particles are transferred to the surface as a result of random turbulent air, impaction, and /or gravitational settling.

Earth Observing System (abbreviated **EOS**)
a major NASA initiative to develop and deploy state-of-the-art *remote sensing* instruments for global studies of the land surface, biosphere, solid earth, atmosphere, oceans, and cryosphere. The first EOS satellite, Terra, was launched in December 1999. Other EOS satellites include Aqua, Aura, ICESat, among others.

Emission of radiation
the generation and sending out of radiant energy. The emission of radiation by natural emitters is accompanied by a loss of energy and is considered separately from the processes of *absorption* or *scattering*.

Emission of gases or particles
the introduction of gaseous or particulate matter into the atmosphere by natural or human activities, e.g., bubble bursting of *whitecaps*, agriculture or wild fires, volcanic eruptions, and industrial processes.

Equilibrium vapor pressure
the pressure of a vapor in equilibrium with its condensed phase (liquid or solid).

Evaporation (also called vaporization)
physical process (phase transition) by which a liquid is transformed to the gaseous state; the opposite of *condensation.*

External mixture (referring to an *aerosol*; contrasted with *internal mixture*)
an aerosol in which different particles (or in some usages, different particles in the same size range) exhibit different compositions.

Extinction (sometimes called attenuation)
the process of removal of radiant energy from an incident beam by the processes of *absorption* and/or *scattering* and consisting of the totality of this removal.

Extinction coefficient
fraction of incident radiant energy removed by extinction per length of travel of radiation through the substance.

General circulation model (abbreviated **GCM**)
a time-dependent numerical model of the entire global atmosphere or ocean or both. The acronym GCM is often applied to Global Climate Model.

Geostationary satellite
a satellite to be placed into a circular orbit in a plane aligned with Earth's equator, and at an altitude of approximately 36,000 km such that the orbital period of the satellite is exactly equal to Earth's period of rotation (approximately 24 hours). The satellite appears stationary with respect to a fixed point on the rotating Earth.

Hygroscopicity
the relative ability of a substance (as an *aerosol*) to adsorb water vapor from its surroundings and ultimately dissolve. Frequently reported as ratio of some property of particle or of particulate phase of an aerosol (e.g., diameter, mean diameter) as function of *relative humidity* to that at low relative humidity.

Ice nucleus (abbreviated **IN**)
any particle that serves as a nucleus leading to the formation of ice crystals without regard to the particular physical processes involved in the nucleation.

In situ
a method of obtaining information about properties of an object (e.g., *aerosol*, cloud) through direct contact with that object, as opposed to *remote sensing*.

Internal mixture (referring to an *aerosol*; contrasted with external mixture)
an aerosol consisting of a mixture of two or more substances, for which all particles exhibit the same composition (or in some usage, the requirement of identical composition is limited to all particles in a given size range). Typically an internal mixture has a higher *absorption coefficient* than an external mixture.

Irradiance (also called radiant flux density)
a radiometric term for the rate at which radiant energy in a radiation field is transferred across a unit area of a surface (real or imaginary) in a hemisphere of directions. In general, irradiance depends on the orientation of the surface. The radiant energy may be confined to a narrow range of frequencies (spectral or monochromatic irradiance) or integrated over a broad range of frequencies.

Large eddy simulation (LES)
A three dimensional numerical simulation of turbulent flow in which large eddies (with scales on the order of hundreds of meters) are resolved and the effects of the subgrid-scale eddies are parameterized. The typical model grid-size is < 100 m and modeling domains are on the order of 10 km. Because they resolve cloud-scale dynamics, large eddy simulations are powerful tools for studying the effects of aerosol on cloud microphysics and dynamics.

Lidar (light detection and ranging)
a technique for detecting and characterizing objects by transmitting pulses of laser light and analyzing the portion of the signal that is reflected and returned to the sensor.

Liquid water path
line integral of the mass concentration of the liquid water droplets in the atmosphere along a specified path, typically along the path above a point on the Earth surface to the top of the atmosphere.

Longwave radiation (also known as terrestrial radiation or thermal infrared radiation)
electromagnetic radiation at wavelengths greater than 4 µm, typically for temperatures characteristic of Earth's surface or atmosphere. In practice, radiation originating by *emission* from Earth and its atmosphere, including clouds; contrasted with *shortwave radiation*.

Low Earth orbit (LEO)
an orbit (of satellite) typically between 300 and 2000 kilometers above Earth.

Mass spectrometer
instrument that fragments and ionizes a chemical substance or mixture by and characterizes composition by amounts of ions as function of molecular weight.

Nucleation
the process of initiation of a new phase in a supercooled (for liquid) or supersaturated (for solution or vapor) environment; the initiation of a phase change of a substance to a lower thermodynamic energy state (vapor to liquid condensation, vapor to solid deposition, liquid to solid freezing).

Optical depth
the *optical thickness* measured vertically above some given altitude. Optical depth is dimensionless and may be applied to Rayleigh scattering optical depth, aerosol *extinction* (or *scattering*, or *absorption*) *optical depth*.

Optical thickness
line integral of *extinction* (or *scattering* or *absorption*) co-*efficient* along a path. Dimensionless.

Passive remote sensing
a remote sensing system that relies on the emission (transmission) of natural levels of radiation from (through) the target. Contrasted with *active remote sensing*.

Phase function
probability distribution function of the angular distribution of the intensity of radiation scattered (by a molecule, gas, particle or aerosol) relative to the direction of the incident beam. See also *Aerosol phase function*.

Polarization
a state in which rays of light exhibit different properties in different directions as measured azimuthially about the direction of propagation of the radiation, especially the state in which all the electromagnetic vibration takes place in a single plane (plane polarization).

Polarimeter
instrument that measures the polarization of incoming light often used in the characterization of light scattered by atmospheric aerosols.

Primary trace atmospheric gases or particles
substances which are directly emitted into the atmosphere from Earth surface, vegetation or natural or human activity, e.g., bubble bursting of *whitecaps*, fires, and industrial processes; contrasted with *secondary* substances.

Radar (radio detection and ranging)
similar to lidar, but using radiation in microwave range.

Radiance
a radiometric term for the rate at which radiant energy in a set of directions confined to a small unit solid angle around a particular direction is transferred across unit area of a surface (real or imaginary) projected onto this direction, per unit solid angle of incident direction.

Radiative forcing
the net energy flux (downwelling minus upwelling) difference between an initial and a perturbed state of atmospheric constituents, such as carbon dioxide or aerosols, at a specified level in the atmosphere; applies also to perturbation in reflected radiation at Earth's surface due to change in albedo. See also *Aerosol radiative forcing*.

Radiative heating
the process by which temperature of an object (or volume of space that encompasses a gas or aerosol) increases in response to an excess of absorbed radiation over emitted radiation.

Radiometer
instrument that measures the intensity of radiant energy radiated by an object at a given wavelength; may or may not resolve by wavelength.

Refractive index (of a medium)
the real part is a measure for how much the speed of light (or other waves such as sound waves) is reduced inside the medium relative to speed of light in vacuum, and the imaginary part is a measure of the amount of *absorption* when the electromagnetic wave propagates through the medium.

Relative humidity
the ratio of the vapor pressure of water to its saturation vapor pressure at the same temperature.

Remote sensing: a method of obtaining information about properties of an object (e.g., aerosol, cloud) without coming into physical contact with that object; opposed to *in situ*.

Saturation
the condition in which the vapor pressure (of a liquid substance; for atmospheric application, water) is equal to the *equilibrium vapor pressure* of the substance over a plane surface of the pure liquid substance, sometimes similarly for ice; similarly for a solute in contact with a solution.

Scattering
in a broad sense, the process by which matter is excited to radiate by an external source of electromagnetic radiation. By this definition, reflection, refraction, and even diffraction of electromagnetic waves are subsumed under *scattering*. Often the term scattered radiation is applied to that radiation observed in directions other than that of the source and may also be applied to acoustic and other waves.

Scattering coefficient
fraction of incident radiant energy removed by *scattering* per length of travel of radiation through the substance.

Secondary trace atmospheric gases or particles
formed in the atmosphere by chemical reaction, new particle formation, etc.; contrasted with *primary* substances, which are directly emitted into the atmosphere.

Secondary organic aerosols (SOA)
organic *aerosol* particles formed in the atmosphere by chemical reactions from gas-phase precursors.

Shortwave radiation
radiation in the visible and near-visible portions of the electromagnetic spectrum (roughly 0.3 to 4.0 µm in wavelength) which range encompasses the great majority of solar radiation and little longwave (terrestrial thermal) radiation; contrasted with *longwave (terrestrial) radiation*.

Single scattering albedo (SSA)
the ratio of light scattering to total light extinction (sum of *scattering* and *absorption*); for *aerosols*, generally restricted to scattering and extinction by the aerosol particles. More absorbing aerosols have lower SSA; a value of unity indicates that the particles are not absorbing.

Solar zenith angle
angle between the vector of Sun and the zenith.

Spectrometer
instrument that measures light received in terms of the intensity at constituent wavelengths, used for example to determine chemical makeup, temperature profiles, and other properties of atmosphere. See also *Mass spectrometer*.

Stratosphere
the region of the atmosphere extending from the top of the *troposphere*, at heights of roughly 10-17 km, to the base of the mesosphere, at a height of roughly 50 km.

Sunglint
a phenomenon that occurs when the sun reflects off the surface of the ocean at the same angle that a satellite sensor is viewing the surface.

Supersaturation
the condition existing in a given portion of the atmosphere (or other space) when the *relative humidity* is greater than 100%, that is, when it contains more water vapor than is needed to produce *saturation* with respect to a plane surface of pure water or pure ice.

Surface albedo
the ratio, often expressed as a percentage, of the amount of electromagnetic radiation reflected by Earth's surface to the amount incident upon it. In general, surface albedo depends on wavelength and the directionality of the incident radiation; hence whether incident radiation is direct or diffuse, cf., *bidirectional reflectance distribution function* (*BRDF*). Value varies with wavelength and with the surface composition. For example, the surface albedo of snow and ice vary from 80% to 90% in the mid-visible, and that of bare ground from 10% to 20%.

Troposphere

the portion of the atmosphere from the earth's surface to the tropopause; that is, the lowest 10-20 kilometers of the atmosphere, depending on latitude and season; most weather occurs in troposphere.

Transient climate response

The time-dependent surface temperature response to a gradually evolving forcing.

Wet scavenging or wet deposition

removal of trace substances from the air by either rain or snow. May refer to in-cloud scavenging, uptake of trace substances into cloud water followed by precipitation, or to below-cloud scavenging, uptake of material below cloud by falling precipitation and subsequent delivery to Earth's surface.

Whitecap

a patch of white water formed at the crest of a wave as it breaks, due to air being mixed into the water.

Major reference: *Glossary of Meteorology*, 2nd edition, American Meteorological Society.

ACRONYMS

A	Surface albedo (broadband)
$\mathring{A}$	Ångström exponent
ABC	Asian Brown Cloud
ACE	Aerosol Characterization Experiment
AD-Net	Asian Dust Network
ADEOS	Advanced Earth Observation Satellite
ADM	Angular Dependence Models
AeroCom	Aerosol Comparisons between Observations and Models
AERONET	Aerosol Robotic Network
AI	Aerosol Index
AIOP	Aerosol Intensive Operative Period
ANL	Argonne National Laboratory (DOE)
AOD (τ)	Aerosol Optical Depth
AOT	Aerosol Optical Thickness
APS	Aerosol Polarimetry Sensor
AR4	Forth Assessment Report, IPCC
ARCTAS	Arctic Research of the Composition of the Troposphere from Aircraft and Satellites
ARM	Atmospheric Radiation Measurements
AVHRR	Advanced Very High Resolution Radiometer
A-Train	Constellation of six afternoon overpass satellites
BASE-A	Biomass Burning Airborne and Space-borne Experiment Amazon and Brazil
BC	Black Carbon
BNL	Brookhaven National Laboratory (DOE)
BRDF	Bidirectional Reflectance Distribution Function
CALIOP	Cloud and Aerosol Lidar with Orthogonal Polarization
CALIPSO	Cloud Aerosol Infrared Pathfinder Satellite Observations
CAPMoN	Canadian Air and Precipitation Monitoring Network
CCN	Cloud Condensation Nuclei
CCRI	Climate Change Research Initiative
CCSP	Climate Change *Science,* Program
CDNC	Cloud Droplet Number Concentration
CERES	Clouds and the Earth's Radiant Energy System
CLAMS	Chesapeake Lighthouse and Aircraft Measurements for Satellite campaign
CTM	Chemistry and Transport Model
DABEX	Dust And Biomass-burning Experiment
DOE	Department of Energy
DRF	Direct Radiative Forcing (aerosol)
EANET	Acid Deposition Monitoring Network in East Asia
EARLINET	European Aerosol Research Lidar Network
EarthCARE	Earth Clouds, Aerosols, and Radiation Explorer

EAST-AIRE	East Asian Studies of Tropospheric Aerosols: An International Regional Experiment		**LMDZ**	Laboratoire de Météorologie Dynamique with Zoom, France
EMEP	European Monitoring and Evaluation Programme		**LOA**	Laboratoire d' Optique Atmosphérique, France
EOS	Earth Observing System		**LOSU**	Level of Scientific Understanding
EP	Earth Pathfinder		**LSCE**	Laboratoire des Sciences du Climat et de l'Environnement, France
EPA	Environmental Protection Agency		**LWC**	Liquid Water Content
ERBE	Earth Radiation Budget Experiment		**LWP**	Liquid Water Path
ESRL	Earth System Research Laboratory (NOAA)		**MAN**	Maritime Aerosol Network
			MEE	Mass Extinction Efficiency
E_τ	Aerosol Forcing Efficiency (RF normalized by AOD)		**MILAGRO**	Megacity Initiative: Local and Global Research Observations
FAR	IPCC First Assessment Report (1990)		**MFRSR**	Multifilter Rotating Shadowband Radiometer
FT	Free Troposphere		**MINOS**	Mediterranean Intensive Oxidant Study
g	Particle scattering asymmetry factor		**MISR**	Multi-angle Imaging SpectroRadiometer
GAW	Global Atmospheric Watch		**MODIS**	Moderate Resolution Imaging Spectro-radiometer
GCM	General Circulation Model, Global Climate Model		**MOZART**	Model for Ozone and Related chemical Tracers
GEOS	Goddard Earth Observing System		**MPLNET**	Micro Pulse Lidar Network
GFDL	Geophysical Fluid Dynamics Laboratory (NOAA)		**NASA**	National Aeronautics and Space Administration
GHGs	Greenhouse Gases		**NASDA**	NAtional Space Development Agency, Japan
GISS	Goddard Institute for Space Studies (NASA)		**NEAQS**	New England Air Quality Study
GLAS	Geoscience Laser Altimeter System		**NOAA**	National Oceanography and Atmosphere Administration
GMI	Global Modeling Initiative		**NPOESS**	National Polar-orbiting Operational Environmental Satellite System
GOCART	Goddard Chemistry Aerosol Radiation and Transport (model)		**NPP**	NPOESS Preparatory Project
GOES	Geostationary Operational Environmental Satellite		**NPS**	National Park Services
GoMACCS	Gulf of Mexico Atmospheric Composition and Climate Study		**NRC**	National Research Council
GSFC	Goddard Space Flight Center (NASA)		**OC**	Organic Carbon
HSRL	High-Spectral-Resolution Lidar		**OMI**	Ozone Monitoring Instrument
ICARTT	International Consortium for Atmospheric Research on Transport and Transformation		**PARASOL**	Polarization and Anisotropy of Reflectance for Atmospheric Science, coupled with Observations from a Lidar
ICESat	Ice, Cloud, and Land Elevation Satellite			
IMPROVE	Interagency Monitoring of Protected Visual Environment		**PDF**	Probability Distribution Function
INCA	Interactions between Chemistry and Aerosol (LMDz model)		**PEM-West**	Western Pacific Exploratory Mission
			PM	Particulate Matter (aerosols)
INDOEX	Indian Ocean Experiment		**PMEL**	Pacific Marine Environmental Laboratory (NOAA)
INTEX-NA	Intercontinental Transport Experiment - North America		**POLDER**	Polarization and Directionality of the Earth's Reflectance
INTEX-B	Intercontinental Transport Experiment - Phase B		**POM**	Particulate Organic Matter
IPCC	Intergovernmental Panel on Climate Change		**PRIDE**	Pueto Rico Dust Experiment
			REALM	Regional East Atmospheric Lidar Mesonet
IR	Infrared radiation		**RF**	Radiative Forcing, aerosol
LBA	Large-Scale Biosphere-Atmosphere Experiment in Amazon		**RH**	Relative Humidity
			RTM	Radiative Transfer Model
LES	Large Eddy Simulation		**SAFARI**	South Africa Regional Science, Experiment
LITE	Lidar In-space Technology Experiment			

SAMUM	Saharan Mineral Dust Experiment	**SZA**	Solar Zenith Angle
SAP	Synthesis and Assessment Product (CCSP)	**TAR**	Third Assessment Report, IPCC
SAR	IPCC Second Assessment Report (1995)	**TARFOX**	Tropospheric Aerosol Radiative Forcing Observational Experiment
SCAR-A	Smoke, Clouds, and Radiation - America		
SCAR-B	Smoke, Clouds, and Radiation - Brazil	**TCR**	Transient Climate sensitivity Range
SeaWiFS	Sea-viewing Wide Field-of-view Sensor	**TexAQS**	Texas Air Quality Study
SGP	Southern Great Plain, ARM site in Oklahoma	**TOA**	Top of the Atmosphere
		TOMS	Total Ozone Mapping Spectrometer
SHADE	Saharan Dust Experiment	**TRACE-A**	Transport and Chemical Evolution over the Atlantic
SMOCC	Smoke, Aerosols, Clouds, Rainfall and Climate		
		TRACE-P	Transport and Chemical Evolution over the Pacific
SOA	Secondary Organic Aerosol		
SPRINTARS	Spectral Radiation-Transport Model for Aerosol Species	**UAE2**	United Arab Emirates Unified Aerosol Experiment
SSA	Single-Scattering Albedo	**UMBC**	University of Maryland at Baltimore County
SST	Sea Surface Temperature		
STEM	Sulfate Transport and Deposition Model	**UV**	Ultraviolet radiation
SURFRAD	NOAA's national surface radiation budget network	**VOC**	Volatile Organic Compounds
		WMO	World Meteorological Organization

Assessing the environmental impact of cloud fields becomes even more complicated when the contributions of aerosol particles in and around the cloud particles are also considered. Image from MODIS. Credit: NASA.

REFERENCES

Abdou, W., D. Diner, J. Martonchik, C. Bruegge, R. Kahn, B. Gaitley, and K. Crean, 2005: Comparison of coincident MISR and MODIS aerosol optical depths over land and ocean scenes containing AERONET sites. *Journal of Geophysical Research*, **110**, D10S07, doi:10.1029/2004JD004693.

Ackerman, A.S., Toon, O. B., and P. V. Hobbs, 1994: Reassessing the dependence of cloud condensation nucleus concentration on formation rate. *Nature*, **367**, 445-447, doi:10.1038/367445a0.

Ackerman, A., O. Toon, D. Stevens, A. Heymsfield, V. Ramanathan, and E. Welton, 2000: Reduction of tropical cloudiness by soot. *Science*, **288**, 1042-1047.

Ackerman, A. S., M. P. Kirkpatrick, D. E. Stevens and O. B. Toon, 2004: The impact of humidity above stratiform clouds on indirect aerosol climate forcing. *Nature*, **432**, 1014-1017.

Ackerman, T., and G. Stokes, 2003: The Atmospheric Radiation Measurement Program. *Physics Today* **56**, 38-44.

Albrecht, B., 1989: Aerosols, cloud microphysics, and fractional cloudiness. *Science*, **245**, 1227-1230.

Alpert, P., P. Kishcha, Y. Kaufman, and R. Schwarzbard, 2005: Global dimming or local dimming? Effect of urbanization on sunlight availability. *Geophysical Research Letters*, **32**, L17802, doi: 10.1029/GL023320.

Anderson, T., R. Charlson, S. Schwartz, R. Knutti, O. Boucher, H. Rodhe, and J. Heintzenberg, 2003: Climate forcing by aerosols—A hazy picture. *Science*, **300**, 1103-1104.

Anderson, T., R. Charlson, N. Bellouin, O. Boucher, M. Chin, S. Christopher, J. Haywood, Y. Kaufman, S. Kinne, J. Ogren, L. Remer, T. Takemura, D. Tanré, O. Torres, C. Trepte, B. Wielicki, D. Winker, and H. Yu, 2005a: An "A-Train" strategy for quantifying direct aerosol forcing of climate. *Bulletin of the American Meteorological Society*, **86**, 1795-1809.

Anderson, T., Y. Wu, D. Chu, B. Schmid, J. Redemann, and O. Dubovik, 2005b: Testing the MODIS satellite retrieval of aerosol fine-mode fraction. *Journal of Geophysical Research*, **110**, D18204, doi:10.1029/2005JD005978.

Andreae, M. O., D. Rosenfeld, P. Artaxo, A. A. Costa, G. P. Frank, K. M. Longo and M. A. F. Silvas-Dias, 2004: Smoking rain clouds over the amazon. *Science*, **303**, 1337-1342.

Andrews, E., P. J. Sheridan, J. A. Ogren, R. Ferrare, 2004: *In situ* aerosol profiles over the Southern Great Plains cloud and radiation test bed site: 1. Aerosol optical properties. *Journal of Geophysical Research*, **109**, D06208, doi:10.1029/2003JD004025.

Ansmann, A., U. Wandinger, A. Wiedensohler, and U. Leiterer, 2002: Lindenderg Aerosol Characterization Experiment 1998 (LACE 98): Overview, *Journal of Geophysical Research*, **107**, 8129, doi:10.1029/2000JD000233.

Arnott, W., H. Moosmuller, and C. Rogers, 1997: Photoacoustic spectrometer for measuring light absorption by aerosol: instrument description. *Atmospheric Environment*, **33**, 2845-2852.

Atwater, M., 1970: Planetary albedo changes due to aerosols. *Science*, **170**(3953), 64-66.

Augustine, J.A., G.B. Hodges, E.G. Dutton, J.J. Michalsky, and C.R. Cornwall, 2008: An aerosol optical depth climatology for NOAA's national surface radiation budget network (SURFRAD). *Journal of Geophysical Research*, **113**, D11204, doi:10.1029/2007JD009504.

Baker, M. B., and R.J. Charlson, 1990: Bistability of CCN concentrations and thermodynamics in the cloud-topped boundary layer. *Nature*, **345**, 142-145.

Balkanski, Y., M. Schulz, T. Claquin, and S. Guibert, 2007: Reevaluation of mineral aerosol radiative forcings suggests a better agreement with satellite and AERONET data. *Atmospheric Chemistry and Physics*, **7**, 81-95.

Bates, T., B. Huebert, J. Gras, F. Griffiths, and P. Durkee (1998): The International Global Atmospheric Chemistry (IGAC) Project's First Aerosol Characterization Experiment (ACE-1)—Overview. *Journal of Geophysical Research*, **103**, 16297-16318.

Bates, T.S., P.K. Quinn, D.J. Coffman, J.E. Johnson, T.L. Miller, D.S. Covert, A. Wiedensohler, S. Leinert, A. Nowak, and C. Neusüb, 2001: Regional physical and chemical properties of the marine boundary layer aerosol across the Atlantic during Aerosols99: An overview. *Journal of Geophysical Research*, **106**, 20767-20782.

Bates T., P. Quinn, D. Coffman, D. Covert, T. Miller, J. Johnson, G. Carmichael, S. uazzotti, D. Sodeman, K. Prather, M. Rivera, L. Russell, and J. Merrill, 2004: Marine boundary layer dust and pollution transport associated with the passage of a frontal system over eastern Asia. *Journal of Geophysical Research*, **109**, doi:10.1029/2003JD004094.

Bates T., et al., 2006: Aerosol direct radiative effects over the northwestern Atlantic, northwestern Pacific, and North Indian Oceans: estimates based on *in situ* chemical and optical measurements and chemical transport modeling. *Atmospheric Chemistry and Physics*, **6**, 1657-1732.

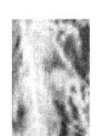

Baynard, T., E.R. Lovejoy, A. Pettersson, S.S. Brown, D. Lack, H. Osthoff, P. Massoli, S. Ciciora, W.P. Dube, and A.R. Ravishankara, 2007: Design and application of a pulsed cavity ring-down aerosol extinction spectrometer for field measurements. *Aerosol Science and Technology,* **41**, 447-462.

Bellouin, N., O. Boucher, D. Tanré, and O. Dubovik, 2003: Aerosol absorption over the clear-sky oceans deduced from POL-DER-1 and AERONET observations. *Geophysical Research Letters,* **30**, 1748, doi:10.1029/2003GL017121.

Bellouin, N., O. Boucher, J. Haywood, and M. Reddy, 2005: Global estimates of aerosol direct radiative forcing from satellite measurements. *Nature,* **438**, 1138-1140, doi:10.1038/nature04348.

Bellouin, N., A. Jones, J. Haywood, and S.A. Christopher, 2008: Updated estimate of aerosol direct radiative forcing from satellite observations and comparison against the Hadley Centre climate model. *Journal of Geophysical Research,* **113**, D10205, doi:10.1029/2007JD009385.

Bond, T.C., D.G. Streets, K.F. Yarber, S.M. Nelson, J.-H. Woo, and Z. Klimont, 2004: A technology-based global inventory of black and organic carbon emissions from combustion. *Journal of Geophysical Research,* **109**, D14203, doi:10.1029/2003JD003697.

Bond, T.C., E. Bhardwaj, R. Dong, R. Jogani, S. Jung, C. Roden, D.G. Streets, and N.M. Trautmann, 2007: Historical emissions of black and organic carbon aerosol from energy-related combustion, 1850-2000. *Global Biogeochemical Cycles,* **21**, GB2018, doi:10.1029/2006GB002840.

Boucher, O., and D. Tanré, 2000: Estimation of the aerosol perturbation to the Earth's radiative budget over oceans using POLDER satellite aerosol retrievals. *Geophysical Research Letters,* **27**, 1103-1106.

Brenguier, J. L., P. Y. Chuang, Y. Fouquart, D. W. Johnson, F. Parol, H. Pawlowska, J. Pelon, L. Schuller, F. Schroder, and J. Snider, 2000: An overview of the ACE-2 CLOUDYCOLUMN closure experiment. *Tellus,* **52B**, 815-827.

Caldeira, K., A. K. Jain, and M. I. Hoffert, 2003: Climate sensitivity uncertainty and the need for energy without CO_2 emission. *Science,* **299**, 2052-2054.

Carmichael, G., G. Calori, H. Hayami, I. Uno, S. Cho, M. Engardt, S. Kim, Y. Ichikawa, Y. Ikeda, J. Woo, H. Ueda and M. Amann, 2002: The Mics-Asia study: Model intercomparison of long-range transport and sulfur deposition in East Asia. *Atmospheric Environment,* **36**, 175-199.

Carmichael, G., Y. Tang, G. Kurata, I. Uno, D. Streets, N. Thongboonchoo, J. Woo, S. Guttikunda, A. White, T. Wang, D. Blake, E. Atlas, A. Fried, B. Potter, M. Avery, G. Sachse, S. Sandholm, Y. Kondo, R. Talbot, A. Bandy, D. Thorton and A. Clarke, 2003: Evaluating regional emission estimates using the TRACE-P observations. *Journal of Geophysical Research,* **108**, 8810, doi:10.1029/2002JD003116.

Carrico, C. et al., 2005: Hygroscopic growth behavior of a carbon-dominated aerosol in Yosemite National Park. *Atmospheric Environment,* **39**, 1393-1404.

CCSP, 2008: *Climate Projections Based on Emissions Scenarios for Long-lived and Short-lived Radiatively Active Gases and Aerosols.* A Report by the U.S. Climate Change Science, Program and the Subcommittee on Global Change Research, H. Levy II, D, T. Shindell, A. Gilliland, M. D. Schwarzkopf, L. W. Horowitz, (eds.). Department of Commerce, NOAA's National Climatic Data Center, Washington, D. C. USA, 116 pp.

Chand, D., T. Anderson, R. Wood, R. J. Charlson, Y. Hu, Z. Liu, and M. Vaughan, 2008: Quantifying above-cloud aerosol using spaceborne lidar for improved understanding of cloudy-sky direct climate forcing. *Journal of Geophysical Research,* **113**, D13206, doi:10.1029/2007JD009433.

Charlson, R. and M. Pilat, 1969: Climate: The influence of aerosols. *Journal of Applied Meteorology,* **8**, 1001-1002.

Charlson, R., J. Langner, and H. Rodhe, 1990: Sulfate aerosol and climate. *Nature,* **348**, 22.

Charlson, R., J. Langner, H. Rodhe, C. Leovy, and S. Warren, 1991: Perturbation of the Northern Hemisphere radiative balance by backscattering from anthropogenic sulfate aerosols. *Tellus,* **43AB**, 152-163.

Charlson, R., S. Schwartz, J. Hales, R. Cess, R. J. Coakley, Jr., J. Hansen, and D. Hofmann, 1992: Climate forcing by anthropogenic aerosols. *Science,* **255**, 423-430.

Chen, W-T, R. Kahn, D. Nelson, K. Yau, and J. Seinfeld, 2008: Sensitivity of multi-angle imaging to optical and microphysical properties of biomass burning aerosols. *Journal of Geophysical Research,* **113**, D10203, doi:10.1029/2007JD009414.

Chin, M., P. Ginoux, S. Kinne, O. Torres, B. Holben, B. Duncan, R. Martin, J. Logan, A. Higurashi, and T. Nakajima, 2002: Tropospheric aerosol optical thickness from the GOCART model and comparisons with satellite and sun photometer measurements. *Journal of the Atmospheric Sciences,* **59**, 461-483.

Chin, M., T. Diehl, P. Ginoux, and W. Malm, 2007: Intercontinental transport of pollution and dust aerosols: implications for regional air quality. *Atmospheric Chemistry and Physics,* **7**, 5501-5517.

Chou, M., P. Chan, and M. Wang, 2002: Aerosol radiative forcing derived from SeaWiFS-retrieved aerosol optical properties. *Journal of the Atmospheric Sciences,* **59**, 748-757.

Christopher, S., and J. Zhang, 2002: Daytime variation of short-wave direct radiative forcing of biomass burning aerosols from GEOS-8 imager. *Journal of the Atmospheric Sciences,* **59**, 681-691.

Christopher, S., J. Zhang, Y. Kaufman, and L. Remer, 2006: Satellite-based assessment of top of atmosphere anthropogenic aerosol radiative forcing over cloud-free oceans. *Geophysical Research Letters,* **33**, L15816.

Christopher, A., and T. Jones, 2008: Short-wave aerosol radiative efficiency over the global oceans derived from satellite data. *Tellus,* (B) **60(4)**, 636-640.

Chu, D., Y. Kaufman, C. Ichoku, L. Remer, D. Tanré, and B. Holben, 2002: Validation of MODIS aerosol optical depth retrieval over land. *Geophysical Research Letters,* **29**, 8007, doi:10.1029/2001/GL013205.

Chung, C., V. Ramanathan, D. Kim, and I. Podgomy, 2005: Global anthropogenic aerosol direct forcing derived from satellite and ground-based observations. *Journal of Geophysical Research,* **110**, D24207, doi:10.1029/2005JD006356.

Chung, C. E. and G. Zhang, 2004: Impact of absorbing aerosol on precipitation. *Journal of Geophysical Research,* **109**, doi:10.1029/2004JD004726.

Clarke, A.D., J.N. Porter, F.P.J. Valero, and P. Pilewskie, 1996: Vertical profiles, aerosol microphysics, and optical closure during the Atlantic Stratocumulus Transition Experiment: Measured and modeled column optical properties *Journal of Geophysical Research,* **101**, 4443-4453.

Coakley, J. Jr., R. Cess, and F. Yurevich, 1983: The effect of tropospheric aerosols on the earth's radiation budget: A parameterization for climate models. *Journal of the Atmospheric Sciences,* **40**, 116-138.

Coakley, J. A. Jr. and C. D. Walsh, 2002: Limits to the aerosol indirect radiative effect derived from observations of ship tracks. *Journal of the Atmospheric Sciences,* **59**, 668-680.

Collins, D.R., H.H. Jonsson, J.H. Seinfeld, R.C. Flagan, S. Gassó, D.A. Hegg, P.B. Russell, B. Schmid, J.M. Livingston, E. Öström, K.J. Noone, L.M. Russell, and J.P. Putaud, 2000: *In Situ* aerosol size distributions and clear column radiative closure during ACE-2. *Tellus,* **52B**, 498-525.

Collins, W., P. Rasch, B. Eaton, B. Khattatov, J. Lamarque, and C. Zender, 2001: Simulating aerosols using a chemical transport model with assimilation of satellite aerosol retrievals: Methodology for INDOEX. *Journal of Geophysical Research,* **106**, 7313-7336.

Conant, W. C., T. M. VanReken, T. A. Rissman, V. Varutbangkul, H. H. Jonsson, A. Nenes, J. L. Jimenez, A. E. Delia, R. Bahreini, G. C. Roberts, R. C. Flagan,J. H. Seinfeld, 2004: Aerosol, cloud drop concentration closure in warm cumulus. *Journal of Geophysical Research,* **109**, D13204, doi:10.1029/2003JD004324.

Cooke, W.F., and J.J.N. Wilson, 1996: A global black carbon aerosol model. *Journal of Geophysical Research,,* **101**, 19395-19409.

Cooke, W.F., C. Liousse, H. Cachier, and J. Feichter, 1999: Construction of a 1° × 1° fossil fuel emission data set for carbonaceous aerosol and implementation and radiative impact in the ECHAM4 model. *Journal of Geophysical Research,* **104**, 22137-22162.

Costa, M., A. Silva, and V. Levizzani, 2004a: Aerosol characterization and direct radiative forcing assessment over the ocean. Part I: Methodology and sensitivity analysis. *Journal of Applied Meteorology,* **43**, 1799-1817.

Costa, M., A. Silva AM, and V. Levizzani, 2004b: Aerosol characterization and direct radiative forcing assessment over the ocean. Part II: Application to test cases and validation. *Journal of Applied Meteorology,* **43**, 1818-1833.

de Gouw, J., et al., 2005: Budget of organic carbon in a polluted atmosphere: Results from the New England Air Quality Study in 2002. *Journal of Geophysical Research,* **110**, D16305, doi:10.1029/2004JD005623.

Delene, D. and J. Ogren, 2002: Variability of aerosol optical properties at four North American surface monitoring sites. *Journal of the Atmospheric Sciences,* **59**, 1135-1150.

Delworth, T. L., V. Ramaswamy and G. L. Stenchikov, 2005: The impact of aerosols on simulated ocean temperature and heat content in the 20th century. *Geophysical Research Letters,* **32**, doi:10.1029/2005GL024457.

Dentener, F., S. Kinne, T. Bond, O. Boucher, J. Cofala, S. Generoso, P. Ginoux, S. Gong, J.J. Hoelzemann, A. Ito, L. Marelli, J.E. Penner, J.-P. Putaud, C. Textor, M. Schulz, G.R. van der Werf, and J. Wilson, 2006: Emissions of primary aerosol and precursor gases in the years 2000 and 1750 prescribed datasets for AeroCom. *Atmospheric Chemistry and Physics,* **6**, 4321-4344.

Deuzé, J., F. Bréon, C. Devaux, P. Goloub, M. Herman, B. Lafrance, F. Maignan, A. Marchand, F. Nadal, G. Perry, and D. Tanré, 2001: Remote sensing of aerosols over land surfaces from POLDER-ADEOS-1 polarized measurements. *Journal of Geophysical Research,* **106**, 4913-4926.

Diner, D., J. Beckert, T. Reilly, et al., 1998: Multiangle Imaging SptectrRadiometer (MISR) description and experiment overview. *IEEE Transactions on Geoscience and Remote Sensing,* **36**, 1072-1087.

Diner, D., J. Beckert, G. Bothwell and J. Rodriguez, 2002: Performance of the MISR instrument during its first 20 months in Earth orbit. *IEEE Transactions on Geoscience and Remote Sensing,* **40**, 1449-1466.

Diner, D., T. Ackerman, T. Anderson, et al., 2004: Progressive Aerosol Retrieval and Assimilation Global Observing Network (PARAGON): An integrated approach for characterizing aerosol climatic and environmental interactions. *Bulletin of the American Meteorological Society,* **85**, 1491-1501.

Doherty, S.J., P. Quinn, A. Jefferson, C. Carrico, T.L. Anderson, and D. Hegg, 2005: A comparison and summary of aerosol optical properties as observed *in situ* from aircraft, ship and land during ACE-Asia. *Journal of Geophysical Research,* **110**, D04201, doi: 10.1029/2004JD004964.

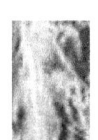

Dubovik, O., A. Smirnov, B. Holben, M. King, Y. Kaufman, and Slutsker, 2000: Accuracy assessments of aerosol optical properties retrieved from AERONET sun and sky radiance measurements. *Journal of Geophysical Research,* **105,** 9791-9806.

Dubovik, O., and M. King, 2000: A flexible inversion algorithm for retrieval of aerosol optical properties from Sun and sky radiance measurements. *Journal of Geophysical Research,,* **105,** 20673-20696.

Dubovik, O., B. Holben, T. Eck, A. Smirnov, Y. Kaufman, M. King, D. Tanré, and I. Slutsker, 2002: Variability of absorption and optical properties of key aerosol types observed in worldwide locations. *Journal of the Atmospheric Sciences,* **59,** 590-608.

Dubovik, O., T. Lapyonok, Y. Kaufman, M. Chin, P. Ginoux, and A. Sinyuk, 2007: Retrieving global sources of aerosols from MODIS observations by inverting GOCART model, *Atmospheric Chemistry and Physics Discussions,* 7, 3629-3718.

Dusek, U., G. P. Frank, L. Hildebrandt, J. Curtius, S. Walter, D. Chand, F. Drewnick, S. Hings, D. Jung, S. Borrmann, and M. O. Andreae, 2006: Size matters more than chemistry in controlling which aerosol particles can nucleate cloud droplets. *Science,* **312,** 1375-1378.

Eagan, R.C., P. V. Hobbs and L. F. Radke, 1974: Measurements of cloud condensation nuclei and cloud droplet size distributions in the vicinity of forest fires. *Journal of Applied Meteorology,* **13,** 553-557.

Eck, T., B. Holben, J. Reid, O. Dubovik, A. Smirnov, N. O'Neill, I. Slutsker, and S. Kinne, 1999: Wavelength dependence of the optical depth of biomass burning, urban and desert dust aerosols. *Journal of Geophysical Research,* **104,** 31333-31350.

Eck, T., et al., 2008: Spatial and temporal variability of column-integrated aerosol optical properties in the southern Arabian Gulf and United Arab Emirates in summer. *Journal of Geophysical Research,* **113,** D01204, doi:10.1029/2007JD008944.

Ervens, B., G. Feingold, and S. M. Kreidenweis, 2005: The influence of water-soluble organic carbon on cloud drop number concentration. *Journal of Geophysical Research,* **110,** D18211, doi:10.1029/2004JD005634.

Fehsenfeld, F., et al., 2006:International Consortium for *Atmospheric Research* on Transport and Transformation (ICARTT): North America to Europe—Overview of the 2004 summer field study. *Journal of Geophysical Research,* **111,** D23S01, doi:10.1029/2006JD007829.

Feingold, G., B. Stevens, W.R. Cotton, and R.L. Walko, 1994: An explicit microphysics/LES model designed to simulate the Twomey Effect. *Atmospheric Research,* **33,** 207-233.

Feingold, G., W. R. Cotton, S. M. Kreidenweis, and J. T. Davis, 1999: The impact of giant cloud condensation nuclei on drizzle formation in stratocumulus: Implications for cloud radiative properties. *Journal of the Atmospheric Sciences,* **56,** 4100-4117.

Feingold, G., Remer, L. A., Ramaprasad, J. and Kaufman, Y. J., 2001: Analysis of smoke impact on clouds in Brazilian biomass burning regions: An extension of Twomey's approach. *Journal of Geophysical Research,* **106,** 22907-22922.

Feingold, G. W. Eberhard, D. Veron, and M. Previdi, 2003: First measurements of the Twomey aerosol indirect effect using ground-based remote sensors. *Geophysical Research Letters,* **30,** 1287, doi:10.1029/2002GL016633.

Feingold, G., 2003: Modeling of the first indirect effect: Analysis of measurement requirements. *Geophysical Research Letters,* **30,** 1997, doi:10.1029/2003GL017967.

Feingold, G., H. Jiang, and J. Harrington, 2005: On smoke suppression of clouds in Amazonia. *Geophysical Research Letters,* **32,** L02804, doi:10.1029/2004GL021369.

Feingold, G., R. Furrer, P. Pilewskie, L. A. Remer, Q. Min, H. Jonsson, 2006: Aerosol indirect effect studies at Southern Great Plains during the May 2003 Intensive Operations Period. *Journal of Geophysical Research,* **111,** D05S14, doi:10.1029/2004JD005648.

Fernandes, S.D., N.M. Trautmann, D.G. Streets, C.A. Roden, and T.C. Bond, 2007: Global biofuel use, 1850-2000. *Global Biogeochemical Cycles,* **21,** GB2019, doi:10.1029/2006GB002836.

Ferrare, R., G. Feingold, S. Ghan, J. Ogren, B. Schmid, S.E. Schwartz, and P. Sheridan, 2006: Preface to special section: Atmospheric Radiation Measurement Program May 2003 Intensive Operations Period examining aerosol properties and radiative influences. *Journal of Geophysical Research,* **111,** D05S01, doi:10.1029/2005JD006908.

Fiebig, M., and J.A. Ogren, 2006: Retrieval and climatology of the aerosol asymmetry parameter in the NOAA aerosol monitoring network. *Journal of Geophysical Research,* **111,** D21204, doi:10.1029/2005JD006545.

Fishman, J., J.M. Hoell, R.D. Bendura, R.J. McNeal, and V. Kirchhoff, 1996: NASA GTE TRACE A experiment (Septemner-October 2002): Overview. *Journal of Geophysical Research,* **101,** 23865-23880.

Fitzgerald, J. W., 1975: Approximation formulas for the equilibrium size of an aerosol particle as a function of its dry size and composition and the ambient relative humidity. *Journal of Applied Meteorology,* **14,** 1044-1049.

Fraser, R. and Y. Kaufman, 1985: The relative importance of aerosol scattering and absorption in Remote Sensing. *Transactions on Geoscience and Remote Sensing,* GE-**23,** 625-633.

Garrett, T., C. Zhao, X. Dong, G. Mace, and P. Hobbs, 2004: Effects of varying aerosol regimes on low-level Arctic stratus. *Geophysical Research Letters*, **31**, L17105, doi:10.1029/2004GL019928.

Garrett, T., and C. Zhao, 2006: Increased Arctic cloud longwave emissivity associated with pollution from mid-latitudes. *Nature*, **440**, 787-789.

Geogdzhayev, I., M. Mishchenko, W. Rossow, B. Cairns, B., and A. Lacis, 2002: Global two-channel AVHRR retrievals of aerosol properties over the ocean for the period of NOAA-9 observations and preliminary retrievals using NOAA-7 and NOAA-11 data. *Journal of the Atmospheric Sciences*, **59**, 262-278.

Ghan, S., and S.E. Schwartz, 2007: Aerosol properties and processes. *Bulletin of the American Meteorological Society*, **88**, 1059-1083.

Gillett, N.P., et al., 2002a: Reconciling two approaches to the detection of anthropogenic influence on climate. *Journal of Climate*, **15**, 326–329.

Gillett, N.P., et al., 2002b: Detecting anthropogenic influence with a multimodel ensemble. *Geophysical Research Letters*, **29**, doi:10.1029/2002GL015836.

Ginoux, P., M. Chin, I. Tegen, J. M. Prospero, B. Holben, O. Dubovik and S.-J. Lin, 2001: Sources and distributions of dust aerosols simulated with the GOCART model. *Journal of Geophysical Research*, **20**, 20255-20273.

Ginoux, P., L. W. Horowitz, V. Ramaswamy, I. V. Geogdzhayev, B. N. Holben, G. Stenchikov and X. tie, 2006: Evaluation of aerosol distribution and optical depth in the Geophysical Fluid Dynamics Laboratory coupled model CM2.1 for present climate. *Journal of Geophysical Research*, **111**, doi:10.1029/2005JD006707.

Golaz, J-C., V. E. Larson, and W. R. Cotton, 2002a: A PDF-based model for boundary layer clouds. Part I: Method and model description. *Journal of the Atmospheric Sciences*, **59**, 3540-3551.

Golaz, J-C., V. E. Larson, and W. R. Cotton, 2002b: A PDF-based model for boundary layer clouds. Part II: Model results. *Journal of the Atmospheric Sciences*, **59**, 3552-3571.

Grabowski, W.W., 2004: An improved framework for superparameterization. *Journal of the Atmospheric Sciences*, **61**, 1940-52.

Grabowski, W.W., X. Wu, and M.W. Moncrieff, 1999: Cloud resolving modeling of tropical cloud systems during Phase III of GATE. Part III: Effects of cloud microphysics. *Journal of the Atmospheric Sciences*, **56**, 2384-2402.

Gregory, J.M., et al., 2002: An observationally based estimate of the climate sensitivity. *Journal of Climate*, **15**, 3117-3121.

Gunn, R. and B. B. Phillips. 1957: An experimental investigation of the effect of air pollution on the initiation of rain. *Journal of Meteorology*, **14**, 272-280.

Han, Q., W. B. Rossow, J. Chou, and R. M. Welch, 1998: Global survey of the relationship of cloud albedo and liquid water path with droplet size using ISCCP. *Journal of Climate*, **11**, 1516-1528.

Han, Q., W.B. Rossow, J. Zeng, and R. Welch, 2002: Three different behaviors of liquid water path of water clouds in aerosol-cloud interactions. *Journal of the Atmospheric Sciences*, **59**, 726-735.

Hansen, J., M. Sato, and R. Ruedy, 1997: Radiative forcing and climate response. *Journal of Geophysical Research*, **102**, 6831-6864.

Hansen, J., et al., 2005: Efficacy of climate forcings. *Journal of Geophysical Research*, **110**, doi:10.1029/2005JD005776, 45pp.

Hansen, J. et al., 2007: Climate simulations for 1880-2003 with GISS model E. *Climate Dynamics*, **29**, 661-696.

Harrison, L., J. Michalsky, and J. Berndt, 1994: Automated multifilter rotating shadowband radiometer: An instrument for optical depth and radiation measurements. *Applied Optics*, **33**, 5118-5125.

Harvey, L.D.D., 2004: Characterizing the annual-mean climatic effect of anthropogenic CO_2 and aerosol emissions in eight coupled atmosphere-ocean GCMs. *Climate Dynamics*, **23**, 569-599.

Haywood, J. M., V. Ramaswamy, and B. J. Soden, 1999: Tropospheric aerosol climate forcing in clear-sky satellite observations over the oceans. *Science*, **283**(5406), 1299-1303.

Haywood, J., and O. Boucher, 2000: Estimates of the direct and indirect radiative forcing due to tropospheric aerosols: A review. *Reviews of Geophysics*, **38**, 513-543.

Haywood, J., P. Francis, S. Osborne, M. Glew, N. Loeb, E. Highwood, D. Tanré, E. Myhre, P. Formenti, and E. Hirst, 2003: Radiative properties and direct radiative effect of Saharan dust measured by the C-130 aircraft during SHADE: 1.Solar spectrum. *Journal of Geophysical Research*, **108**, 8577, doi:10.1029/2002JD002687.

Haywood, J., and M. Schulz, 2007: Causes of the reduction in uncertainty in the anthropogenic radiative forcing of climate between IPCC (2001) and IPCC (2007). *Geophysical Research Letters*, **34**, L20701, doi:10.1029/2007GL030749.

Haywood, J., et al., 2008: Overview of the Dust and Biomass burning Experiment and African Monsoon Multidisciplinary Analysis Special Observing Period-0. *Journal of Geophysical Research*, **113**, D00C17, doi:10.1029/2008JD010077.

Heald, C. L., D. J. Jacob, R. J. Park, L. M. Russell, B. J. Huebert, J. H. Seinfeld, H. Liao, and R. J. Weber, 2005: A large organic aerosol source in the free troposphere missing from current models. *Geophysical Research Letters*, **32**, L18809, doi:10.1029/2005GL023831.

Heintzenberg, J., et al., 2009: The SAMUM-1 experiment over Southern Morocco: Overview and introduction. *Tellus,* **61B**, in press.

Henze, D. K. and J.H. Seinfeld, 2006: Global secondary organic aerosol from isoprene oxidation. *Geophysical Research Letters,* **33**, L09812, doi:10.1029/2006GL025976.

Herman, J., P. Bhartia, O. Torres, C. Hsu, C. Seftor, and E. Celarier, 1997: Global distribution of UV-absorbing aerosols from Nimbus-7/TOMS data. *Journal of Geophysical Research,* 102, 16911-16922.

Hoell, J.M., D.D. Davis, S.C. Liu, R. Newell, M. Shipham, H. Akimoto, R.J. McNeal, R.J. Bemdura, and J.W. Drewry, 1996: Pacific Exploratory Mission-West A (PEM-WEST A): September-October, 1991. *Journal of Geophysical Research,* **101**, 1641-1653.

Hoell, J.M., D.D. Davis, S.C. Liu, R. Newell, M. Shipham, H. Akimoto, R.J. McNeal, R.J. Bemdura, and J.W. Drewry, 1997: The Pacific Exploratory Mission-West Phase B: February-March, 1994. *Journal of Geophysical Research,* **102**, 28223-28239.

Hoff, R. et al., 2002: Regional East Atmospheric Lidar Mesonet: REALM, in *Lidar Remote Sensing in Atmospheric and Earth Sciences,* edited by L. Bissonette, G. Roy, and G. Vallée, pp. 281-284, Def. R&D Can. Valcartier, Val-Bélair, Que.

Hoff, R., J. Engel-Cox, N. Krotkov, S. Palm, R. Rogers, K. Mc-Cann, L. Sparling, N. Jordan, O. Torres, and J. Spinhirne, 2004: Long-range transport observations of two large forest fire plumes to the northeastern U.S., in *22nd International Laser Radar Conference, ESA Spec. Publ.,* SP-**561**, 683-686.

Holben, B., T. Eck, I. Slutsker, et al., 1998: AERONET—A federated instrument network and data archive for aerosol characterization. *Remote Sensing of the Environment,* **66**, 1-16.

Holben, B., D. Tanré, A. Smirnov, et al., 2001: An emerging ground-based aerosol climatology: aerosol optical depth from AERONET. *Journal of Geophysical Research,* **106**, 12067-12098.

Horowitz, L. W., et al., 2003: A global simulation of tropospheric ozone and related tracers: Description and evaluation of MOZART, version 2. *Journal of Geophysical Research,* **108**, 4784, doi:10.1029/2002JD002853.

Horowitz, L., 2006: Past, present, and future concentrations of tropospheric ozone and aerosols: Methodology, ozone evaluation, and sensitivity to aerosol wet removal. *Journal of Geophysical Research,* **111**, D22211, doi:10.1029/2005JD006937.

Hoyt, D., and C. Frohlich, 1983: Atmospheric transmission at Davos, Switzerland 1909-1979. *Climatic Change,* **5**, 61-71.

Hsu, N., S. Tsay, M. King, and J. Herman, 2004: Aerosol properties over bright-reflecting source regions. *IEEE Transactions on Geoscience and Remote Sensing,* **42**, 557-569.

Huebert, B., T. Bates, P. Russell, G. Shi, Y. Kim, K. Kawamura, G. Carmichael, and T. Nakajima, 2003: An overview of ACE-Asia: strategies for quantifying the relationships between Asian aerosols and their climatic impacts. *Journal of Geophysical Research,* **108**, 8633, doi:10.1029/2003JD003550.

Huneeus, N., and O. Boucher, 2007: One-dimensional variational retrieval of aerosol extinction coefficient from synthetic LIDAR and radiometric measurements. *Journal of Geophysical Research,* **112**, D14303, doi:10.1029/2006JD007625.

Husar, R., J. Prospero, and L. Stowe, 1997: Characterization of tropospheric aerosols over the oceans with the NOAA advanced very high resolution radiometer optical thickness operational product. *Journal of Geophysical Research,* **102**, 16889-16909.

IPCC, 1992: *Climate Change 1992: The Supplementary Report to the IPCC Scientific Assessment.* J. T. Houghton, B. A. Callander and S. K. Varney (eds). Cambridge University Press, Cambridge, UK, 198 pp.

IPCC (Intergovernmental Panel on Climate Change), 1995: *Radiative forcing of climate change and an evaluation of the IPCC IS92 emission scenarios, in Climate Change 1994,* Cambridge Univ. Press, New York, Cambridge University Press, 1995.

IPCC (Intergovernmental Panel on Climate Change), 1996: *Radiative forcing of climate change, in Climate Change 1995,* Cambridge Univ. Press, New York, Cambridge University Press, 1996.

IPCC (Intergovernmental Panel on Climate Change), 2001: *Radiative forcing of climate change, in Climate Change 2001,* Cambridge Univ. Press, New York, Cambridge University Press, 2001.

IPCC (Intergovernmental Panel on Climate Change), 2007: *Changes in Atmospheric Constituents and in Radiative forcing, in Climate Change 2007,* Cambridge University Press, New York, Cambridge University Press, 2007.

Ito, A., and J.E. Penne, 2005: Historical estimates of carbonaceous aerosols from biomass and fossil fuel burning for the period 1870-2000. *Global Biogeochemical Cycles,* **19**, GB2028, doi:10.1029/2004GB002374.

Jacob, D., J. Crawford, M. Kleb, V. Connors, R.J. Bendura, J. Raper, G. Sachse, J. Gille, L. Emmons, and C. Heald, 2003: The Transport and Chemical Evolution over the Pacific (TRACE-P) aircraft mission: design, execution, and first results. *Journal of Geophysical Research,* **108**, 9000, 10.1029/2002JD003276.

Jayne, J. T., D. C. Leard, X. Zhang, P. Davidovits, K. A. Smith, C. E. Kolb, and D. R. Worsnop, 2000: Development of an aerosol mass spectrometer for size and composition analysis of submicron particles. *Aerosol Science and Technology,* **33**, 49-70.

Jeong, M., Z. Li, D. Chu, and S. Tsay, 2005: Quality and Compatibility Analyses of Global Aerosol Products Derived from the Advanced Very High Resolution Radiometer and Moderate Resolution Imaging Spectroradiometer. *Journal of Geophysical Research,* **110**, D10S09, doi:10.1029/2004JD004648.

Jiang, H., and G. Feingold, 2006: Effect of aerosol on warm convective clouds: Aerosol-cloud-surface flux feedbacks in a new coupled large eddy model. *Journal of Geophysical Research,* **111**, D01202, doi:10.1029/2005JD006138.

Jiang, H., H. Xue, A. Teller, G. Feingold, and Z. Levin, 2006: Aerosol effects on the lifetime of shallow cumulus. *Geophysical Research Letters,* **33**, doi: **10**.1029/2006GL026024.

Jiang, H., G. Feingold, H. H. Jonsson, M.-L. Lu, P. Y. Chuang, R. C. Flagan, J. H. Seinfeld, 2008: Statistical comparison of properties of simulated and observed cumulus clouds in the vicinity of Houston during the Gulf of Mexico Atmospheric Composition and Climate Study (GoMACCS). *Journal of Geophysical Research,* **113**, D13205, doi:10.1029/2007JD009304.

Johnson, D. B., 1982: The role of giant and ultragiant aerosol particles in warm rain initiation. *Journal of the Atmospheric Sciences,* **39**, 448-460.

Jones, G.S., et al., 2005: Sensitivity of global scale attribution results to inclusion of climatic response to black carbon. *Geophysical Research Letters,* **32**:L14701, doi:10.1029/2005GL023370.

Junker, C., and C. Liousse, 2008: A global emission inventory of carbonaceous aerosol from historic records of fossil fuel and biofuel consumption for the period 1860-1997. *Atmospheric Chemistry and Physics,* **8**, 1195-1207.

Kahn, R., P. Banerjee, D. McDonald, and D. Diner, 1998: Sensitivity of multiangle imaging to aerosol optical depth, and to pure-particle size distribution and composition over ocean. *Journal of Geophysical Research,* **103**, 32195-32213.

Kahn, R., P. Banerjee, and D. McDonald, 2001: The sensitivity of multiangle imaging to natural mixtures of aerosols over ocean. *Journal of Geophysical Research,* **106**, 18219-18238.

Kahn, R., J. Ogren, T. Ackerman, et al., 2004: Aerosol data sources and their roles within PARAGON. *Bulletin of the American Meteorological Society,* **85**, 1511-1522.

Kahn, R., R. Gaitley, J. Martonchik, D. Diner, K. Crean, and B. Holben, 2005a: MISR global aerosol optical depth validation based on two years of coincident AERONET observations. *Journal of Geophysical Research,* **110**, D10S04, doi:10.1029/2004JD004706.

Kahn, R., W. Li, J. Martonchik, C. Bruegge, D. Diner, B. Gaitley, W. Abdou, O. Dubovik, B. Holben, A. Smirnov, Z. Jin, and D. Clark, 2005b: MISR low-light-level calibration, and implications for aerosol retrieval over dark water. *Journal of the Atmospheric Sciences,* **62**, 1032-1052.

Kahn, R., W. Li, C. Moroney, D. Diner, J. Martonchik, and E. Fishbein, 2007a: Aerosol source plume physical characteristics from space-based multiangle imaging. *Journal of Geophysical Research,* **112**, D11205, doi:10.1029/2006JD007647.

Kahn, R., et al., 2007b: Satellite-derived aerosol optical depth over dark water from MISR and MODIS: Comparisons with AERONET and implications for climatological studies. *Journal of Geophysical Research,* **112**, D18205, doi:10.1029/2006JD008175.

Kalashnikova, O., and R. Kahn, 2006: Ability of multiangle remote sensing observations to identify and distinguish mineral dust types: Part 2. Sensitivity over dark water. *Journal of Geophysical Research,* **111**:D11207, doi:10.1029/2005JD006756.

Kapustin, V.N., A.D. Clarke, Y. Shinozuka, S. Howell, V. Brekhovskikh, T. Nakajima, and A. Higurashi, 2006: On the determination of a cloud condensation nuclei from satellite: Challenges and possibilities. *Journal of Geophysical Research,* **111**, D04202, doi:10.1029/2004JD005527.

Kaufman, Y., 1987: Satellite sensing of aerosol absorption. *Journal of Geophysical Research,* **92**, 4307-4317.

Kaufman, Y.J., A. Setzer, D. Ward, D. Tanre, B. N. Holben, P. Menzel, M. C. Pereira, and R. Rasmussen, 1992: Biomass Burning Airborne and Spaceborne Experiment in the Amazonas (BASE-A). *Journal of Geophysical Research,* **97**, 14581-14599.

Kaufman, Y. J. and Nakajima, T., 1993: Effect of Amazon smoke on cloud microphysics and albedo—Analysis from satellite imagery. *Journal of Applied Meteorology,* **32**, 729-744.

Kaufman, Y. and R. Fraser, 1997: The effect of smoke particles on clouds and climate forcing. *Science,* **277**, 1636-1639.

Kaufman, Y., D. Tanré, L. Remer, E. Vermote, A. Chu, and B. Holben, 1997: Operational remote sensing of tropospheric aerosol over land from EOS moderate resolution imaging spectroradiometer. *Journal of Geophysical Research,* **102**, 17051-17067.

Kaufman, Y.J., P. V. Hobbs, V. W. J. H. Kirchhoff, P. Artaxo, L. A. Remer, B. N. Holben, M. D. King, D. E. Ward, E. M. Prins, K. M. Longo, L. F. Mattos, C. A. Nobre, J. D. Spinhirne, Q. Ji, A. M. Thompson, J. F. Gleason, and S. A. Christopher, 1998: Smoke, clouds, and radiation—Brazil (SCAR-B) experiment. *Journal of Geophysical Research,* **103**, 31783-31808.

Kaufman, Y., D. Tanré, and O. Boucher, 2002a: A satellite view of aerosols in the climate system. *Nature,* **419**, doi:10.1038/nature01091.

Kaufman, Y., J. Martins, L. Remer, M. Schoeberl, and M. Yamasoe, 2002b: Satellite retrieval of aerosol absorption over the oceans using sunglint. *Geophysical Research Letters,* **29**, 1928, doi:10.1029/2002GL015403.

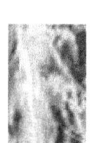

Kaufman, Y., J. Haywood, P. Hobbs, W. Hart, R. Kleidman, and B. Schmid, 2003: Remote sensing of vertical distributions of smoke aerosol off the coast of Africa. *Geophysical Research Letters*, **30**, 1831, doi:10.1029/2003GL017068.

Kaufman, Y., O. Boucher, D. Tanré, M. Chin, L. Remer, and T. Takemura, 2005a: Aerosol anthropogenic component estimated from satellite data. *Geophysical Research Letters*, **32**, L17804, doi:10.1029/2005GL023125.

Kaufman, Y., L. Remer, D. Tanré, R. Li, R. Kleidman, S. Mattoo, R. Levy, T. Eck, B. Holben, C. Ichoku, J. Martins, and I. Koren, 2005b: A critical examination of the residual cloud contamination and diurnal sampling effects on MODIS estimates of aerosol over ocean. *IEEE Transactions on Geoscience and Remote Sensing* **43**, 2886-2897.

Kaufman, Y. J., I. Koren, L. A. Remer, D. Rosenfeld and Y. Rudich, 2005c: The effect of smoke, dust, and pollution aerosol on shallow cloud development over the Atlantic Ocean. *Proceedings of the National Academy of Sciences*, **102**, 11207-11212.

Kaufman, Y. J. and Koren, I., 2006: Smoke and pollution aerosol effect on cloud cover. *Science*, **313**, 655-658.

Kerr, R., 2007: Another global warming icon comes under attack. *Science*, **317**, 28.

Kiehl, J. T., 2007: Twentieth century climate model response and climate sensitivity. *Geophysical Research Letters*, **34**, doi:10.1029/2007GL031383.

Kim, B.-G., S. Schwartz, M. Miller, and Q. Min, 2003: Effective radius of cloud droplets by ground-based remote sensing: Relationship to aerosol. *Journal of Geophysical Research*, **108**, 4740, doi:10.1029/2003JD003721.

Kim, B.-G., M. A. Miller, S. E. Schwartz, Y. Liu, and Q. Min, 2008: The role of adiabaticity in the aerosol first indirect effect. *Journal of Geophysical Research*, **113**, D05210, doi:10.1029/2007JD008961.

Kim, M.-K., K.-M. Lau, M. Chin, K.-M. Kim, Y. Sud, and G. K. Walker, 2006: Atmospheric teleconnection over Eurasia induced by aerosol radiative forcing during boreal spring. *Proceedings of the National Academy of Sciences*, **19**, 4700-4718.

King, M., Y. Kaufman, D. Tanré, and T. Nakajima, 1999: Remote sensing of tropospheric aerosols: Past, present, and future. *Bulletin of the American Meteorological Society*, **80**, 2229-2259.

King, M., S. Platnick, C. Moeller, Revercomb, and D. Chu, 2003: Remote sensing of smoke, land, and clouds from the NASA ER-2 during SAFARI 2000. *Journal of Geophysical Research*, **108**, 8502, doi:10.1029/2002JD003207.

Kinne, S., M. Schulz, C. Textor, et al., 2006: An AeroCom initial assessment—optical properties in aerosol component modules of global models. *Atmospheric Chemistry and Physics*, **6**, 1815-1834.

Kirchstetter, T.W., R.A. Harley, N.M. Kreisberg, M.R. Stolzenburg, and S.V. Hering, 1999: On-road measurement of fine particle and nitrogen oxide emissions from light- and heavy-duty motor vehicles. *Atmospheric Environment*, **33**, 2955-2968.

Kristjánsson, J. E., Stjern, C. W., Stordal, F., Fjæraa, A. M., Myhre, G., and Jónasson, K., 2008: Cosmic rays, cloud condensation nuclei and clouds—a reassessment using MODIS data, *Atmospheric Chemistry and Physics*, **8**, 7373-7387.

Kleinman, L.I. et al., 2008: The time evolution of aerosol composition over the Mexico City plateau. *Atmospheric Chemistry and Physics*, **8**, 1559-1575.

Kleidman, R., N. O'Neill, L. Remer, Y. Kaufman, T. Eck, D. Tanré, O. Dubovik, and B. Holben, 2005: Comparison of Moderate Resolution Imaging Spectroradiometer (MODIS) and Aerosol Robotic Network (AERONET) remote-sensing retrievals of aerosol fine mode fraction over ocean. *Journal of Geophysical Research*, **110**, D22205, doi:10.1029/2005JD005760.

Knutti, R., T.F. Stocker, F. Joos, and G.-K. Plattner, 2002: Constraints on radiative forcing and future climate change from observations and climate model ensembles. *Nature*, **416**, 719-723.

Knutti, R., T.F. Stocker, F. Joos, and G.-K. Plattner, 2003: Probabilistic climate change projections using neural networks. *Climate Dynamics*, **21**, 257-272.

Koch, D., and J. Hansen, 2005: Distant origins of Arctic black carbon: A Goddard Institute for Space Studies ModelE experiment. *Journal of Geophysical Research*, **110**, D04204, doi:10.1029/2004JD005296.

Koch, D., G. Schmidt, and C. Field, 2006: Sulfur, sea salt and radionuclide aerosols in GISS ModelE. *Journal of Geophysical Research*, **111**, D06206, doi:10.1029/2004JD005550.

Koch, D., T.C. Bond, D. Streets, N. Unger, G.R. van der Werf, 2007: Global impact of aerosols from particular source regions and sectors, *Journal of Geophysical Research*, **112**, D02205, doi:10.1029/2005JD007024.

Kogan, Y. L., D. K. Lilly, Z. N. Kogan, and V. Filyushkin, 1994: The effect of CCN regeneration on the evolution of stratocumulus cloud layers. *Atmospheric Research*, **33**, 137-150.

Koren, I., Y. Kaufman, L. Remer, and J. Martins, 2004: Measurement of the effect of Amazon smoke on inhibition of cloud formation. *Science*, **303**, 1342.

Koren, I., Y.J. Kaufman, D. Rosenfeld, L.A. Remer, and Y. Rudich, 2005: Aerosol invigoration and restructuring of Atlantic convective clouds. *Geophysical Research Letters*, **32**, doi:10.1029/2005GL023187.

Koren, I., L.A. Remer, and K. Longo, 2007a: Reversal of trend of biomass burning in the Amazon. *Geophysical Research Letters*, **34**, L20404, doi:10.1029/2007GL031530.

Koren, I., L.A. Remer, Y.J. Kaufman, Y. Rudich, and J.V. Martins, 2007b: On the twilight zone between clouds and aerosols. *Geophysical Research Letters*, **34**, L08805, doi:10.1029/2007GL029253.

Koren, I., J. V. Martins, L. A. Remer, and H. Afargan, 2008: Smoke invigoration versus inhibition of clouds over the Amazon. *Science*, **321**, 946, doi: 10.1126/science.1159185.

Kroll, J. H., N.L. Ng, S.M. Murphy, R.C. Flagan, and J.H. Seinfeld, 2006: Secondary organic aerosol formation from isoprene photooxidation. *Environmental Science and Technology*, **40**, 1869-1877.

Kruger, O. and H. Grasl, 2002: The indirect aerosol effect over Europe. *Geophysical Research Letters*, **29**, doi:10.1029/2001GL014081.

Lack, D., E. Lovejoy, T. Baynard, A. Pettersson, and A. Ravishankara, 2006: Aerosol absorption measurements using photoacoustic spectroscopy: sensitivity, calibration, and uncertainty developments. *Aerosol Science and Technology*, **40**, 697-708.

Larson, V. E., R. Wood, P. R. Field, J.-C. Golaz, T. H. Vonder Haar, and W. R. Cotton, 2001: Small-scale and mesoscale variability of scalars in cloudy boundary layers: One-dimensional probability density functions. *Journal of the Atmospheric Sciences*, **58**, 1978-1996.

Larson, V.E., J.-C. Golaz, H. Jiang and W.R. Cotton, 2005: Supplying local microphysics parameterizations with information about subgrid variability: Latin hypercube sampling. *Journal of the Atmospheric Sciences*, **62**, 4010-4026.

Lau, K., M. Kim, and K. Kim, 2006: Asian summer monsoon anomalies induced by aerosol direct forcing—the role of the Tibetan Plateau. *Climate Dynamics*, **36**, 855-864, doi:10.1007/s00382-006-10114-z.

Lau, K.-M., and K.-M. Kim, 2006: Observational relationships between aerosol and Asian monsoon rainfall, and circulation. *Geophysical Research Letters*, **33**, L21810, doi:10.1029/2006GL027546.

Lau, K.-M., K.-M. Kim, G. Walker, and Y. C. Sud, 2008: A GCM study of the possible impacts of Saharan dust heating on the water cycle and climate of the tropical Atlantic and Caribbean regions. *Proceedings of the National Academy of Sciences*, (submitted).

Leahy, L., T. Anderson, T. Eck, and R. Bergstrom, 2007: A synthesis of single scattering albedo of biomass burning aerosol over southern Africa during SAFARI 2000. *Geophysical Research Letters*, **34**, L12814, doi:10.1029/2007GL029697.

Leaitch, W. R., G.A. Isaac, J.W. Strapp, C.M. Banic and H.A. Wiebe, 1992: The Relationship between Cloud Droplet Number Concentrations and Anthropogenic Pollution—Observations and Climatic Implications. *Journal of Geophysical Research*, **97**, 2463-2474.

Leaitch, W. R., C. M. Banic, G. A. Isaac, M. D. Couture, P. S. K. Liu, I. Gultepe, S.-M. Li, L. Kleinman, J. I. MacPherson, and P. H. Daum, 1996: Physical and chemical observations in marine stratus during the 1993 North Atlantic Regional Experiment: Factors controlling cloud droplet number concentrations. *Journal of Geophysical Research*, **101**, 29123-29135.

Lee, T., et al., 2006: The NPOESS VIIRS day/night visible sensor. *Bulletin of the American Meteorological Society*, **87**, 191-199.

Lelievel, J., H. Berresheim, S. Borrmann, S., et al., 2002: Global air pollution crossroads over the Mediterranean. *Science*, **298**, 794-799.

Léon, J., D. Tanré, J. Pelon, Y. Kaufman, J. Haywood, and B. Chatenet, 2003: Profiling of a Saharan dust outbreak based on a synergy between active and passive remote sensing. *Journal of Geophysical Research*, **108**, 8575, doi:10.1029/2002JD002774.

Levin, Z. and W. R. Cotton, 2008: *Aerosol pollution impact on precipitation: A scientific review.* Report from the WMO/IUGG International Aerosol Precipitation Science, Assessment Group (IAPSAG), World Meteorological Organization, Geneva, Switzerland, 482 pp.

Levy, R., L. Remer, and O. Dubovik, 2007a: Global aerosol optical properties and application to MODIS aerosol retrieval over land. *Journal of Geophysical Research*, **112**, D13210, doi:10.1029/2006JD007815.

Levy, R., L. Remer, S. Mattoo, E. Vermote, and Y. Kaufman, 2007b: Second-generation algorithm for retrieving aerosol properties over land from MODIS spectral reflectance. *Journal of Geophysical Research*, **112**, D13211, doi:10.1029/2006JD007811.

Lewis, E.R. and S.E. Schwartz, 2004: *Sea Salt Aerosol Production: Mechanisms, Methods, Measurements, and Models—A Critical Review.* Geophysical Monograph Series Vol. **152**, (American Geophysical Union, Washington, 2004), 413 pp. ISBN: 0-87590-417-3.

Li, R., Y. Kaufman, W. Hao, I. Salmon, and B. Gao, 2004: A technique for detecting burn scars using MODIS data. *IEEE Transactions on Geoscience and Remote Sensing*, **42**, 1300-1308.

Li, Z., et al., 2007: Preface to special section on East Asian studies of tropospheric aerosols: An international regional experiment (EAST-AIRE). *Journal of Geophysical Research*, **112**, D22s00, doi:10.0129/2007JD008853.

Lindesay, J. A., M.O. Andreae, J.G. Goldammer, G. Harris, H.J. Annegarn, M. Garstang, R.J. Scholes, and B.W. van Wilgen, 1996: International Geosphere Biosphere Programme/International Global Atmospheric Chemistry SAFARI-92 field experiment: Background and overview. *Journal of Geophysical Research*, **101**, 23521-23530.

Liou, K. N. and S-C. Ou, 1989: The Role of Cloud Microphysical Processes in Climate: An Assessment From a One-Dimensional Perspective. *Journal of Geophysical Research*, **94**, 8599-8607.

Liousse, C., J. E. Penner, C. Chuang, J. J. Walton, H. Eddleman and H. Cachier, 1996: A three-dimensional model study of carbonaceous aerosols. *Journal of Geophysical Research,* **101,** 19411-19432.

Liu, H., R. Pinker, and B. Holben, 2005: A global view of aerosols from merged transport models, satellite, and ground observations. *Journal of Geophysical Research,* **110,** D10S15, doi:10.1029/2004JD004695.

Liu, L., A. A. Lacis, B. E. Carlson, M. I. Mishchenko, and B. Cairns, 2006: Assessing Goddard Institute for Space Studies ModelE aerosol climatology using satellite and ground-based measurements: A comparison study. *Journal of Geophysical Research,* **111,** doi:10.1029/2006JD007334.

Liu, X., J. Penner, B. Das, D. Bergmann, J. Rodriguez, S. Strahan, M. Wang, and Y. Feng, 2007: Uncertainties in global aerosol simulations: Assessment using three meteorological data sets. *Journal of Geophysical Research,* **112,** D11212, doi: 10.1029/2006JD008216.

Liu, Z., A. Omar, M. Vaughan, J. Hair, C. Kittaka, Y. Hu, K. Powell, C. Trepte, D. Winker, C. Hostetler, R. Ferrare, and R. Pierce, 2008: CALIPSO lidar observations of the optical properties of Saharan dust: A case study of long-range transport. *Journal of Geophysical Research,* **113,** D07207, doi:10.1029/2007JD008878.

Lockwood, M., and C. Frohlich, 2007: Recent oppositely directed trends in solar climate forcings and the global mean surface air temperature. *Proceedings of the Royal Society A,* 1-14, doi:10.1098/rspa.2007.1880.

Loeb, N., and S. Kato, 2002: Top-of-atmosphere direct radiative effect of aerosols over the tropical oceans from the Clouds and the Earth's Radiant Energy System (CERES) satellite instrument. *Proceedings of the National Academy of Sciences,* **15,** 1474-1484.

Loeb, N., and N. Manalo-Smith, 2005: Top-of-Atmosphere direct radiative effect of aerosols over global oceans from merged CERES and MODIS observations. *Journal of Climate,* **18,** 3506-3526.

Loeb, N. G., S. Kato, K. Loukachine, and N. M. Smith, 2005: Angular distribution models for top-of-atmosphere radiative flux estimation from the Clouds and the Earth's Radiant Energy System instrument on the Terra Satellite. part I: Methodology. *Journal of Atmospheric and Oceanic Technology,* **22,** 338–351.

Lohmann, U., J. Feichter, C. C. Chuang, and J. E. Penner, 1999: Prediction of the number of cloud droplets in the ECHAM GCM. *Journal of Geophysical Research,* **104,** 9169-9198.

Lohmann, U., et al., 2001: Vertical distributions of sulfur species simulated by large scale atmospheric models in COSAM: Comparison with observations. *Tellus,* **53B,** 646-672.

Lohmann, U. and J. Feichter, 2005: Global indirect aerosol effects: a review. *Atmospheric Chemistry and Physics,* **5,** 715-737.

Lohmann, U., I. Koren and Y.J. Kaufman, 2006: Disentangling the role of microphysical and dynamical effects in determining cloud properties over the Atlantic. *Geophysical Research Letters,* **33,** L09802, doi:10.1029/2005GL024625.

Lu, M.-L., G. Feingold, H. Jonsson, P. Chuang, H. Gates, R. C. Flagan, J. H. Seinfeld, 2008: Aerosol-cloud relationships in continental shallow cumulus. *Journal of Geophysical Research,* **113,** D15201, doi:10.1029/2007JD009354.

Lubin, D., S. Satheesh, G. McFarquar, and A. Heymsfield, 2002: Longwave radiative forcing of Indian Ocean tropospheric aerosol. *Journal of Geophysical Research,* **107,** 8004, doi:10.1029/2001JD001183.

Lubin, D. and A. Vogelmann, 2006: A climatologically significant aerosol longwave indirect effect in the Arctic. *Nature,* **439,** 453-456.

Luo, Y., D. Lu, X. Zhou, W. Li, and Q. He, 2001: Characteristics of the spatial distribution and yearly variation of aerosol optical depth over China in last 30 years. *Journal of Geophysical Research,* **106,** 14501, doi:10.1029/2001JD900030.

Magi, B., P. Hobbs, T. Kirchstetter, T. Novakov, D. Hegg, S. Gao, J. Redemann, and B. Schmid, 2005: Aerosol properties and chemical apportionment of aerosol optical depth at locations off the United States East Coast in July and August 2001. *Journal of the Atmospheric Sciences,* **62,** 919-933.

Malm, W., J. Sisler, D. Huffman, R. Eldred, and T. Cahill, 1994: Spatial and seasonal trends in particle concentration and optical extinction in the United States. *Journal of Geophysical Research,* **99,** 1347-1370.

Martins, J., D. Tanré, L. Remer, Y. Kaufman, S. Mattoo, and R. Levy, 2002: MODIS cloud screening for remote sensing of aerosol over oceans using spatial variability. *Geophysical Research Letters,* **29,** 10.1029/2001GL013252.

Martonchik, J., D. Diner, R. Kahn, M. Verstraete, B. Pinty, H. Gordon, and T. Ackerman, 1998a: Techniques for the Retrieval of aerosol properties over land and ocean using multiangle data. *IEEE Transactions on Geoscience and Remote Sensing,* **36,** 1212-1227

Martonchik, J., D. Diner, B. Pinty, M. Verstraete, R. Myneni, Y. Knjazikhin, and H. Gordon, 1998b: Determination of land and ocean reflective, radiative, and biophysical properties using multiangle imaging. *IEEE Transactions on Geoscience and Remote Sensing,* **36,** 1266-1281.

Martonchik, J., D. Diner, K. Crean, and M. Bull, 2002: Regional aerosol retrieval results from MISR. *IEEE Transactions on Geoscience and Remote Sensing,* **40,** 1520-1531.

Massie, S., O. Torres, and S. Smith, 2004: Total ozone mapping spectrometer (TOMS) observations of increases in Asian aerosol in winter from 1979 to 2000. *Journal of Geophysical Research,* **109,** D18211, doi:10.1029/2004JD004620.

Matheson, M. A., J. A. Coakley Jr., W. R. Tahnk, 2005: Aerosol and cloud property relationships for summertime stratiform clouds in the northeastern Atlantic from Advanced Very High Resolution Radiometer observations. *Journal of Geophysical Research,* **110**, D24204, doi:10.1029/2005JD006165.

Matsui, T., and R. Pielke, Sr., 2006: Measurement-based estimation of the spatial gradient of aerosol radiative forcing. *Geophysical Research Letters,* **33**, L11813, doi:10.1029/2006GL025974.

Matsui, T., H. Masunaga, S. M. Kreidenweis, R. A. Pielke Sr., W.-K. Tao, M. Chin, Y. J. Kaufman, 2006: Satellite-based assessment of marine low cloud variability associated with aerosol, atmospheric stability, and the diurnal cycle. *Journal of Geophysical Research,* **111**, D17204, doi:10.1029/2005JD006097.

Matthis, I., A. Ansmann, D. Müller, U. Wandinger, and D. Althausen, 2004: Multiyear aerosol observations with dual-wavelength Raman lidar in the framework of EARLINET. *Journal of Geophysical Research,* **109**, D13203, doi:10.1029/2004JD004600.

McComiskey, A., and G. Feingold, 2008: Quantifying error in the radiative forcing of the first aerosol indirect effect, *Geophysical Research Letters,* **35**, L02810, doi:10.1029/2007GL032667.

McComiskey, A., S.E. Schwartz, B. Schmid, H. Guan, E.R. Lewis, P. Ricchiazzi, and J.A. Ogren, 2008a: Direct aerosol forcing: Calculation from observables and sensitivity to inputs. *Journal of Geophysical Research,* **113**, D09202, doi:10.1029/2007JD009170.

McComiskey, A, G. Feingold, A. S. Frisch, D. Turner, M. Miller, J. C. Chiu, Q. Min, and J. Ogren, 2008b: An assessment of aerosol-cloud interactions in marine stratus clouds based on surface remote sensing. *Journal of Geophysical Research,* submitted.

McCormick, R., and J. Ludwig, 1967: Climate modification by atmospheric aerosols. *Science,* **156**, 1358-1359.

McCormick, M. P., L. W. Thomason, and C. R. Trepte 1995: Atmospheric effects of the Mt. Pinatubo eruption. *Nature,* **373**, 399-404.

McFiggans, G., P. Artaxo, U. Baltensberger, H. Coe, M.C. Facchini, G. Feingold, S. Fuzzi, M. Gysel, A. Laaksonen, U. Lohmann, T. F. Mentel, D. M. Murphy, C. D. O'Dowd, J. R. Snider, E. Weingartner, 2006: The effect of physical and chemical aerosol properties on warm cloud droplet activation. *Atmospheric Chemistry and Physics,* **6**, 2593-2649.

Menon, S., A.D. Del Genio, Y. Kaufman, R. Bennartz, D. Koch, N. Loeb, and D. Orlikowski, 2008: Analyzing signatures of aerosol-cloud interactions from satellite retrievals and the GISS GCM to constrain the aerosol indirect effect. *Journal of Geophysical Research,* **113**, D14S22, doi:10.1029/2007JD009442.

Michalsky, J., J. Schlemmer, W. Berkheiser, et al., 2001: Multi-year measurements of aerosol optical depth in the Atmospheric Radiation Measurement and Quantitative Links program. *Journal of Geophysical Research,* **106**, 12099-12108.

Min, Q., and L.C. Harrison, 1996: Cloud properties derived from surface MFRSR measurements and comparison with GEOS results at the ARM SGP site. *Geophysical Research Letters,* **23**, 1641- 1644.

Minnis P., E. F. Harrison, L. L. Stowe, G. G. Gibson, F. M. Denn, D. R. Doelling. and W. L. Smith. Jr., 1993: Radiative climate forcing by the Mount Pinatubo eruption. *Science,* **259**, 411-1415.

Mishchenko, M., I. Geogdzhayev, B. Cairns, W. Rossow, and A. Lacis, 1999: Aerosol retrievals over the ocean by use of channels 1 and 2 AVHRR data: Sensitivity analysis and preliminary results. *Applied Optics,* **38**, 7325-7341.

Mishchenko, M., et al., 2007a: Long-term satellite record reveals likely recent aerosol trend. *Science,* **315**, 1543.

Mishchenko, M., et al., 2007b: Accurate monitoring of terrestrial aerosols and total solar irradiance. *Bulletin of the American Meteorological Society,* **88**, 677-691.

Mishchenko, M., and I. V. Geogdzhayev, 2007: Satellite remote sensing reveals regional tropospheric aerosol trends. *Optics Express,* **15**, 7423-7438.

Mitchell, J. Jr., 1971: The effect of atmospheric aerosols on climate with special reference to temperature near the Earth's surface. *Journal of Applied Meteorology,* **10**, 703-714.

Molina, L. T., S. Madronich, J.S. Gaffney, and H.B. Singh, 2008: Overview of MILAGRO/INTEX-B Campaign. IGAC activities, *Newsletter of International Global Atmospheric Chemistry Project* **38**, 2-15, April, 2008.

Moody, E., M. King, S. Platnick, C. Schaaf, and F. Gao, 2005: Spatially complete global spectral surface albedos: value-added datasets derived from Terra MODIS land products. *IEEE Transactions on Geoscience and Remote Sensing,* **43**, 144-158.

Mouillot, F., A. Narasimha, Y. Balkanski, J.-F. Lamarque, and C.B. Field, 2006: Global carbon emissions from biomass burning in the 20th century. *Geophysical Research Letters,* **33**, L01801, doi:10.1029/2005GL024707.

Murayama, T., N. Sugimoto, I. Uno, I., et al., 2001: Ground-based network observation of Asian dust events of April 1998 in East Asia. *Journal of Geophysical Research,* **106**, 18346-18359.

NRC (National Research Council), 2001: *Climate Change Sciences: An analysis of some key questions,* 42pp., National Academy Press, Washington D.C..

NRC (National Research Council), 2005: *Radiative Forcing of Climate Change: Expanding the Concept and Addressing Uncertainties,* National Academy Press, Washington D.C. (Available at http://www.nap.edu/openbook/0309095069/html).

Nakajima, T., Higurashi, A., Kawamoto, K. and Penner, J. E., 2001: A possible correlation between satellite-derived cloud and aerosol microphysical parameters. *Geophysical Research Letters*, **28**, 1171-1174.

Norris, J., and M. Wild, 2007: Trends in aerosol radiative effects over Europe inferred from observed cloud cover, solar "dimming", and solar "brightening". *Journal of Geophysical Research*, **112**, D08214, doi:10.1029/2006JD007794.

Novakov, T., V. Ramanathan, J. Hansen, T. Kirchstetter, M. Sato, J. Sinton, and J. Sathaye, 2003: Large historical changes of fossil-fuel black carbon emissions. *Geophysical Research Letters*, **30**, 1324, doi:10.1029/2002GL016345.

O'Dowd, C. D., et al. 1999: The relative importance of sea-salt and nss-sulphate aerosol to the marine CCN population: An improved multi-component aerosol-droplet parameterization. *Quarterly Journal of the Royal Meteorological Society*, **125**, 1295-1313.

O'Neill, N., T. Eck, A. Smirnov, B. Holben, and S. Thulasiraman, 2003: Spectral discrimination of coarse and fine mode optical depth. *Journal of Geophysical Research*, **108**(D17), 4559, doi:10.1029/2002JD002975.

Patadia, F., P. Gupta, and S.A. Christopher, 2008: First observational estimates of global clear-sky shortwave aerosol direct radiative effect over land. *Journal of Geophysical Research*, **35**, L04810, doi:10.0129/2007GL032314.

Penner, J., R. Dickinson, and C. O'Neill, 1992: Effects of aerosol from biomass burning on the global radiation budget. *Science*, **256**, 1432-1434.

Penner, J., R. Charlson, J. Hales, et al., 1994: Quantifying and minimizing uncertainty of climate forcing by anthropogenic aerosols, *Bulletin of the American Meteorological Society*, **75**, 375-400.

Penner, J.E., H. Eddleman, and T. Novakov, 1993: Towards the development of a global inventory for black carbon emissions. *Atmospheric Environment*, **27**, 1277-1295.

Penner, J. E. et al., 2002: A comparison of model- and satellite-derived aerosol optical depth and reflectivity. *Journal of the Atmospheric Sciences*, **59**, 441-460.

Penner, J. E., et al. 2006: Model intercomparison of indirect aerosol effects. *Atmospheric Chemistry and Physics*, **6**, 3391-3405.

Pincus, R., and S.A. Klein, 2000: Unresolved spatial variability and microphysical process rates in large-scale models. *Journal of Geophysical Research*, **105**, 27059-27065.

Pinker, R., B. Zhang, and E. Dutton, 2005: Do satellites detect trends in surface solar radiation? *Science*, **308**, 850-854.

Procopio, A. S., P. Artaxo, Y. J. Kaufman, L. A. Remer, J. S. Schafer, and B. N. Holben, 2004: Multiyear analysis of Amazonian biomass burning smoke radiative forcing of climate. *Journal of Geophysical Research*, **31**, L03108, doi: 10.1029/2003GL018646.

Qian, Y., W. Wang, L Leung, and D. Kaiser, 2007: Variability of solar radiation under cloud-free skies in China: The role of aerosols. *Geophysical Research Letters*, **34**, L12804, doi:10.1029/2006GL028800.

Quaas, J., and O. Boucher, 2005: Constraining the first aerosol indirect radiative forcing in the LMDZ GCM using POLDER and MODIS satellite data. *Geophysical Research Letters*, **32**, L17814.

Quaas, J., O. Boucher and U. Lohmann, 2006: Constraining the total aerosol indirect effect in the LMDZ GCM and ECHAM4 GCMs using MODIS satellite data. *Atmospheric Chemistry and Physics Discussions*, **5**, 9669-9690.

Quaas, J., O. Boucher, N. Bellouin, and S. Kinne, 2008: Satellite-based estimate of the direct and indirect aerosol climate forcing. *Journal of Geophysical Research*, **113**, D05204, doi:10.1029/2007JD008962.

Quinn, P.K., T. Anderson, T. Bates, R. Dlugi, J. Heintzenberg, W. Von Hoyningen-Huene, M. Kumula, P. Russel, and E. Swietlicki, 1996: Closure in tropospheric aerosol-climate research: A review and future needs for addressing aerosol direct shortwave radiative forcing. *Contributions to Atmospheric Physics*, **69**, 547-577.

Quinn, P.K., D. Coffman, V. Kapustin, T.S. Bates and D.S. Covert, 1998: Aerosol optical properties in the marine boundary layer during ACE 1 and the underlying chemical and physical aerosol properties. *Journal of Geophysical Research*, **103**, 16547-16563.

Quinn P.K., T. Bates, T. Miller, D. Coffman, J. Johnson, J. Harris, J. Ogren, G. Forbes, G., T. Anderson, D. Covert, and M. Rood, 2000: Surface submicron aerosol chemical composition: What fraction is not sulfate? *Journal of Geophysical Research*, **105**, 6785-6806.

Quinn, P.K., T.L. Miller, T.S. Bates, J.A. Ogren, E. Andrews, and G.E. Shaw, 2002: A three-year record of simultaneously measured aerosol chemical and optical properties at Barrow, Alaska. *Journal of Geophysical Research*, **107**(D11), doi:10.1029/2001JD001248.

Quinn, P.K., and T. Bates, 2003: North American, Asian, and Indian haze: Similar regional impacts on climate? *Geophysical Research Letters*, **30**, 1555, doi:10.1029/2003GL016934.

Quinn, P.K., D.J. Coffman, T.S. Bates, E.J. Welton, D.S. Covert, T.L. Miller, J.E. Johnson, S. Maria, L. Russell, R. Arimoto, C.M. Carrico, M.J. Rood, and J. Anderson, 2004: Aerosol optical properties measured aboard the Ronald H. Brown during ACE-Asia as a function of aerosol chemical composition and source region. *Journal of Geophysical Research*, **109**, doi:10.1029/2003JD004010.

Quinn, P.K. and T. Bates, 2005: Regional Aerosol Properties: Comparisons from ACE 1, ACE 2, Aerosols99, INDOEX, ACE Asia, TARFOX, and NEAQS. *Journal of Geophysical Research*, **110**, D14202, doi:10.1029/2004JD004755.

Quinn, P.K., et al., 2005: Impact of particulate organic matter on the relative humidity dependence of light scattering: A simplified parameterization. *Geophysical Research Letters*, **32**, L22809, doi:101029/2005GL024322.

Quinn, P.K., G. Shaw, E. Andrews, E.G. Dutton, T. Ruoho-Airola, S.L. Gong, 2007: Arctic Haze: Current trends and knowledge gaps. *Tellus*, **59B**, 99-114.

Radke, L.F., J.A. Coakley Jr., and M.D. King, 1989: Direct and remote sensing observations of the effects of ship tracks on clouds. *Science*, **246**, 1146-1149.

Raes, F., T. Bates, F. McGovern, and M. van Liedekerke, 2000: The 2nd Aerosol Characterization Experiment (ACE-2): General overview and main results. *Tellus*, **52B**, 111-125.

Ramanathan, V., P. Crutzen, J. Kiehl, and D. Rosenfeld, 2001a: Aerosols, Climate, and the Hydrological Cycle. *Science*, **294**, 2119-2124.

Ramanathan, V., P. Crutzen, J. Lelieveld, et al., 2001b: Indian Ocean Experiment: An integrated analysis of the climate forcing and effects of the great Indo-Asian haze. *Journal of Geophysical Research*, **106**, 28371-28398.

Ramanathan, V., and P. Crutzen, 2003: Atmospheric Brown "Clouds". *Atmospheric Environment*, **37**, 4033-4035.

Ramanathan, V., et al., 2005: Atmospheric brown clouds: Impact on South Asian climate and hydrologic cycle. *Proceedings of the National Academy of Sciences*, USA, **102**, 5326-5333.

Randall, D., M. Khairoutdinov, A. Arakawa, and W. Grabowski, 2003: Breaking the cloud parameterization deadlock. *Bulletin of the American Meteorological Society*, **84**, 1547-1564.

Rao, S., K. Riahi, K. Kupiainen, and Z. Klimont, 2005: Long-term scenarios for black and organic carbon emissions. *Environmental Science*, **2**, 205-216.

Reddy, M., O. Boucher, N. Bellouin, M. Schulz, Y. Balkanski, J. Dufresne, and M. Pham, 2005a: Estimates of multi-component aerosol optical depth and direct radiative perturbation in the LMDZT general circulation model. *Journal of Geophysical Research*, **110**, D10S16, doi:10.1029/2004JD004757.

Reddy, M., O. Boucher, Y. Balkanski, and M. Schulz, 2005b: Aerosol optical depths and direct radiative perturbations by species and source type. *Geophysical Research Letters*, **32**, L12803, doi:10.1029/2004GL021743.

Reid, J., J. Kinney, and D. Wesphal, et al., 2003: Analysis of measurements of Saharan dust by airborne and ground-based remote sensing methods during the Puerto Rico Dust Experiment (PRIDE). *Journal of Geophysical Research*, **108**, 8586, doi:10.1029/2002JD002493.

Reid, J., et al., 2008: An overview of UAE2 flight operations: Observations of summertime atmospheric thermodynamic and aerosol profiles of the southern Arabian Gulf. *Journal of Geophysical Research*, **113**, D14213, doi:10.1029/2007JD009435.

Remer, L., S. Gassó, D. Hegg, Y. Kaufman, and B. Holben, 1997: Urban/industrial aerosol: ground based sun/sky radiometer and airborne *in situ* measurements. *Journal of Geophysical Research*, **102**, 16849-16859.

Remer, L., D. Tanré, Y. Kaufman, C. Ichoku, S. Mattoo, R. Levy, D. Chu, B. Holben, O. Dubovik, A. Smirnov, J. Martins, R. Li, and Z. Ahman, 2002: Validation of MODIS aerosol retrieval over ocean. *Geophysical Research Letters*, **29**, 8008, doi:10.1029/2001/GL013204.

Remer, L., Y. Kaufman, D. Tanré, S. Mattoo, D. Chu, J. Martins, R. Li, C. Ichoku, R. Levy, R. Kleidman, T. Eck, E. Vermote, and B. Holben, 2005: The MODIS aerosol algorithm, products and validation. *Journal of the Atmospheric Sciences*, **62**, 947-973.

Remer, L., and Y. Kaufman, 2006: Aerosol direct radiative effect at the top of the atmosphere over cloud free ocean derived from four years of MODIS data. *Atmospheric Chemistry and Physics*, **6**, 237-253.

Remer, L., et al., 2008: An emerging aerosol climatology from the MODIS satellite sensors, *Journal of Geophysical Research*, **113**, D14S01, doi:10.1029/2007JD009661.

Rissler, J., E. Swietlicki, J. Zhou, G. Roberts, M. O. Andreae, L. V. Gatti, and P. Artaxo 2004: Physical properties of the sub-micrometer aerosol over the Amazon rain forest during the wet-to-dry season transition—comparison of modeled and measured CCN concentrations. *Atmospheric Chemistry and Physics*, **4**, 2119-2143.

Robock, A., 2000: Volcanic eruptions and climate. *Reviews of Geophysics*, **38**(2), 191-219.

Robock, A., 2002: Pinatubo eruption: The climatic aftermath. *Science*, **295**, 1242-1244.

Roderick, M. L. and G. D. Farquhar, 2002: The cause of decreased pan evaporation over the past 50 years. *Science*, **298**, 1410-1411.

Rosenfeld, D., and I. Lansky, 1998: Satellite-based insights into precipitation formation processes in continental and maritime convective clouds. *Bulletin of the American Meteorological Society*, **79**, 2457-2476.

Rosenfeld, D., 2000: Suppression of rain and snow by urban and industrial air pollution. *Science*, **287**, 1793-1796.

Rosenfeld, D., 2006: Aerosols, clouds, and climate. *Science*, **312**, 10.1126/science.1128972.

Ruckstuhl, C., et al., 2008: Aerosol and cloud effects on solar brightening and recent rapid warming. *Geophysical Research Letters*, **35**, L12708, doi:10.1029/2008GL034228.

Russell, P., S. Kinne, and R. Bergstrom, 1997: Aerosol climate effects: local radiative forcing and column closure experiments. *Journal of Geophysical Research*, **102**, 9397-9407.

Russell, P., J. Livingston, P. Hignett, S. Kinne, J. Wong, A. Chien, R. Bergstrom, P. Durkee, and P. Hobbs, 1999: Aerosol-induced radiative flux changes off the United States mid-Atlantic coast: comparison of values calculated from sun photometer and *in situ* data with those measured by airborne pyranometer. *Journal of Geophysical Research*, **104**, 2289-2307.

Saxena, P., L. Hildemann, P. McMurry, and J. Seinfeld, 1995: Organics alter hygroscopic behavior of atmospheric particles. *Journal of Geophysical Research*, **100**, 18755-18770.

Schmid, B., J.M. Livingston, P.B. Russell, P.A. Durkee, H.H. Jonsson, D.R. Collins, R.C. Flagan, J.H. Seinfeld, S. Gasso, D.A. Hegg, E. Ostrom, K.J. Noone, E.J. Welton, K.J. Voss, H.R. Gordon, P. Formenti, and M.O. Andreae, 2000: Clearsky closure studies of lower tropospheric aerosol and water vapor during ACE-2 using airborne sunphotometer, airborne *in situ*, space-borne, and ground-based measurements. *Tellus*, **52**, 568-593.

Schmid, B., R. Ferrare, C. Flynn, et al., 2006: How well do state-of-the-art techniques measuring the vertical profile of tropospheric aerosol extinction compare? *Journal of Geophysical Research*, **111**, doi:10.1029/2005JD005837, 2006.

Schmidt, G. A., et al., 2006: Present-day atmospheric simulations using GISS Model E: Comparison to *in situ*, satellite and reanalysis data. *Journal of Climate*, **19**, 153-192.

Schulz, M., C. Textor, S. Kinne, et al., 2006: Radiative forcing by aerosols as derived from the AeroCom present-day and preindustrial simulations. *Atmospheric Chemistry and Physics*, **6**, 5225-5246.

Schwartz, S. E., R. J. Charlson and H. Rodhe, 2007: Quantifying climate change—too rosy a picture? *Nature Reports Climate Change* **2**, 23-24.

Sekiguchi, M., T. Nakajima, K. Suzuki, et al., A study of the direct and indirect effects of aerosols using global satellite data sets of aerosol and cloud parameters. *Journal of Geophysical Research*, **108**, D22, 4699, doi:10.1029/2002JD003359, 2003

Seinfeld, J.H., et al., 1996. *A Plan for a Research Program on Aerosol Radiative Forcing and Climate Change*. National Research Council. 161 pp.

Seinfeld, J. H., G.R. Carmichael, R. Arimoto, et al. 2004: ACE-Asia: Regional climatic and atmospheric chemical effects of Asian dust and pollution. *Bulletin of the American Meteorological Society*, **85**, 367-380.

Sheridan, P., and J. Ogren, 1999: Observations of the vertical and regional variability of aerosol optical properties over central and eastern North America. *Journal of Geophysical Research*, **104**, 16793-16805.

Shindell, D.T., M. Chin, F. Dentener, et al., 2008a: A multi-model assessment of pollution transport to the Arctic. *Atmospheric Chemistry and Physics*, 8, 5353-5372.

Shindell, D.T., H. Levy, II, M.D. Schwarzkopf, L.W. Horowitz, J.-F. Lamarque, and G. Faluvegi, 2008b: Multimodel projections of climate change from short-lived emissions due to human activities. *Journal of Geophysical Research*, **113**, D11109, doi:10.1029/2007JD009152.

Singh, H.B., W.H. Brune, J.H. Crawford, F. Flocke, and D.J. Jacob, 2008: Chemistry and Transport of Pollution over the Gulf of Mexico and the Pacific: Spring 2006 INTEX-B Campaign Overview and First Results. *Atmospheric Chemistry and Physics Discussions*, submitted.

Sinyuk, A., O. Dubovik, B. Holben, T. F. Eck, F.-M. Breon, J. Martonchik, R. A. Kahn, D. Diner, E. F. Vermote, Y. J. Kaurman, J. C. Roger, T. Lapyonok, and I. Slutsker, 2007: Simultaneous retrieval of aerosol and surface properties from a combination of AERONET and satellite data. *Remote Sensing of the Environment*, **107**, 90-108, doi: 10.1016/j.rse.2006.07.022.

Smirnov, A., B. Holben, T. Eck, O. Dubovik, and I. Slutsker, 2000: Cloud screening and quality control algorithms for the AERONET database. *Remote Sensing of the Environment*, **73**, 337-349.

Smirnov, A., B. Holben, T. Eck, I. Slutsker, B. Chatenet, and R. Pinker, 2002: Diurnal variability of aerosol optical depth observed at AERONET (Aerosol Robotic Network) sites. *Geophysical Research Letters*, **29**, 2115, doi:10.1029/2002GL016305.

Smirnov, A., B. Holben, S. Sakerin, et al., 2006: Ship-based aerosol optical depth measurements in the Atlantic Ocean, comparison with satellite retrievals and GOCART model. *Geophysical Research Letters*, **33**, L14817, doi: 10.1029/2006GL026051.

Smith Jr., W.L., et al., 2005: EOS Terra aerosol and radiative flux validation: An overview of the Chesapeake Lighthouse and aircraft measurements from satellites (CLAMS) experiment. *Journal of the Atmospheric Sciences*, **62**, 903-918.

Sokolik, I., D. Winker, G. Bergametti, et al., 2001: Introduction to special section: outstanding problems in quantifying the radiative impacts of mineral dust. *Journal of Geophysical Research*, **106**, 18015-18027.

Sotiropoulou, R.E.P, A. Nenes, P.J. Adams, and J.H. Seinfeld, 2007: Cloud condensation nuclei prediction error from application of Kohler theory: Importance for the aerosol indirect effect. *Journal of Geophysical Research*, **112**, D12202, doi:10.1029/2006JD007834.

Sotiropoulou, R.E.P, J. Medina, and A. Nenes, 2006: CCN predictions: is theory sufficient for assessments of the indirect effect? *Geophysical Research Letters*, **33**, L05816, doi:10.1029/2005GL025148

Spinhirne, J., S. Palm, W. Hart, D. Hlavka, and E. Welton, 2005: Cloud and Aerosol Measurements from the GLAS Space Borne Lidar: initial results. *Geophysical Research Letters*, **32**, L22S03, doi:10.1029/2005GL023507.

Squires, P., 1958: The microstructure and colloidal stability of warm clouds. I. The relation between structure and stability. *Tellus*, **10**, 256-271.

Stanhill, G., and S. Cohen, 2001: Global dimming: a review of the evidence for a widespread and significant reduction in global radiation with discussion of its probable causes and possible agricultural consequences. *Agricultural and Forest Meteorology*, **107**, 255-278.

Stephens, G., D. Vane, R. Boain, G. Mace, K. Sassen, Z. Wang, A. Illingworth, E. O'Conner, W. Rossow, S. Durden, S. Miller, R. Austin, A. Benedetti, and C. Mitrescu, 2002: The CloudSat mission and the A-Train. *Bulletin of the American Meteorological Society*, **83**, 1771-1790.

Stephens, G. L. and J. M. Haynes, 2007: Near global observations of the warm rain coalescence process. *Geophysical Research Letters*, **34**, L20805, doi:10.1029/2007GL030259.

Stern, D.I., 2005: Global sulfur emissions from 1850 to 2000. *Chemosphere*, **58**, 163-175.

Stevens, B., G. Feingold, R. L. Walko and W. R. Cotton, 1996: On elements of the microphysical structure of numerically simulated non-precipitating stratocumulus. *Journal of the Atmospheric Sciences*, **53**, 980-1006.

Storlevmo, T., J.E. Kristjansson, G. Myhre, M. Johnsud, and F. Stordal, 2006: Combined observational and modeling based study of the aerosol indirect effect. *Atmospheric Chemistry and Physics*, **6**, 3583-3601.

Stott, P.A., et al., 2006: Observational constraints on past attributable warming and predictions of future global warming. *Journal of Climate*, **19**, 3055-3069.

Strawa, A., R. Castaneda, T. Owano, P. Baer, and B. Paldus, 2002: The measurement of aerosol optical properties using continuous wave cavity ring-down techniques. *Journal of Atmospheric and Oceanic Technology,* **20**, 454-465.

Streets, D., T. Bond, T. Lee, and C. Jang, 2004: On the future of carbonaceous aerosol emissions. *Journal of Geophysical Research*, **109**, D24212, doi:10.1029/2004JD004902.

Streets, D., and K. Aunan, 2005: The importance of China's household sector for black carbon emissions. *Geophysical Research Letters*, **32**, L12708, doi:10.1029/2005GL022960.

Streets, D., Y. Wu, and M. Chin, 2006a: Two-decadal aerosol trends as a likely explanation of the global dimming/brightening transition. *Geophysical Research Letters*, **33**, L15806, doi:10.1029/2006GL026471.

Streets, D., Q. Zhang, L. Wang, K. He, J. Hao, Y. Tang, and G. Carmichael, 2006b: Revisiting China's CO emissions after TRACE-P: Synthesis of inventories, atmospheric modeling and observations *Journal of Geophysical Research*, **111**, D14306, doi:10.1029/2006JD007118.

Svensmark, H. and E. Friis-Christensen, 1997: Variation of cosmic ray flux and global cloud coverage—a missing link in solar-climate relationships. *Journal of Atmospheric and Solar-Terrestrial Physics*, **59**, 1225-1232.

Takemura, T., T. Nakajima, O. Dubovik, B. Holben, and S. Kinne, 2002: Single-scattering albedo and radiative forcing of various aerosol species with a global three-dimensional model. *Proceedings of the National Academy of Sciences*, **15**, 333-352.

Takemura, T., T. Nozawa,S. Emori, T. Nakajima, and T. Nakajima, 2005: Simulation of climate response to aerosol direct and indirect effects with aerosol transport-radiation model. *Journal of Geophysical Research*, **110**, D02202, doi:10.1029/2004JD005029.

Tang, Y., G. Carmichael, I. Uno, J. Woo, G. Kurata, B. Lefer, R. Shetter, H. Huang, B. Anderson, M. Avery, A. Clarke and D. Blake, 2003: Influences of biomass burning during the Transport and Chemical Evolution Over the Pacific (TRACE-P) experiment identified by the regional chemical transport model. *Journal of Geophysical Research*, **108**, 8824, doi:10.1029/2002JD003110.

Tang, Y., G. Carmichael, J. Seinfeld, D. Dabdub, R. Weber, B. Huebert, A. Clarke, S. Guazzotti, D. Sodeman, K. Prather, I. Uno, J. Woo, D. Streets, P. Quinn, J. Johnson, C. Song, A. Sandu, R. Talbot and J. Dibb, 2004: Three-dimensional simulations of inorganic aerosol distributions in East Asia during spring 2001. *Journal of Geophysical Research*, **109**, D19S23, doi:10.1029/2003JD004201.

Tanré, D., Y. Kaufman, M. Herman, and S. Mattoo, 1997: Remote sensing of aerosol properties over oceans using the MODIS/EOS spectral radiances. *Journal of Geophysical Research*, **102**, 16971-16988.

Tanré, D., J. Haywood, J. Pelon, J. Léon, B. Chatenet, P. Formenti, P. Francis, P. Goloub, E. Highwood, and G. Myhre, 2003: Measurement and modeling of the Saharan dust radiative impact: Overview of the Saharan Dust Experiment (SHADE). *Journal of Geophysical Research*, **108**, 8574, doi:10.1029/2002JD003273.

Textor, C., M. Schulz, S. Guibert, et al., 2006: Analysis and quantification of the diversities of aerosol life cycles within AEROCOM. *Atmospheric Chemistry and Physics*, **6**, 1777-1813.

Textor, C., et al., 2007: The effect of harmonized emissions on aerosol properties in global models—an AeroCom experiment. *Atmospheric Chemistry and Physics*, **7**, 4489-4501.

Tie, X. et al., 2005: Assessment of the global impact of aerosols on tropospheric oxidants. *Journal of Geophysical Research*, **110**, doi:10.1029/2004JD005359.

Torres, O., P. Bhartia, J. Herman, Z. Ahmad, and J. Gleason, 1998: Derivation of aerosol properties from satellite measurements of backscattered ultraviolet radiation: Theoretical bases. *Journal of Geophysical Research*, **103**, 17009-17110.

Torres, O., P. Bhartia, J. Herman, A. Sinyuk, P. Ginoux, and B. Holben, 2002: A long-term record of aerosol optical depth from TOMS observations and comparison to AERONET measurements. *Journal of the Atmospheric Sciences*, **59**, 398-413.

Torres, O., P. Bhartia, A. Sinyuk, E. Welton, and B. Holben, 2005: Total Ozone Mapping Spectrometer measurements of aerosol absorption from space: Comparison to SAFARI 2000 ground-based observations. *Journal of Geophysical Research*, **110**, D10S18, doi:10.1029/2004JD004611.

Turco, R.P., O.B. Toon, R.C. Whitten, J.B. Pollack, and P. Hamill, 1983: The global cycle of particulate elemental carbon: a theoretical assessment, in *Precipitation Scavenging, Dry Deposition, and Resuspension*, ed. H.R. Pruppacher et al., pp. 1337-1351, Elsevier Science, New York.

Twomey, S., 1977: The influence of pollution on the shortwave albedo of clouds. *Journal of the Atmospheric Sciences*, 34, 1149-1152.

van Ardenne, J. A., F.J. Dentener, J. Olivier, J. Klein, C.G.M. Goldewijk, and J. Lelieveld, 2001: A 1° x 1° resolution data set of historical anthropogenic trace gas emissions for the period 1890–1990. *Global Biogeochemical Cycles*, **15**, 909-928.

Veihelmann, B., P. F. Levelt, P. Stammes, and J. P. Veefkind, 2007: Simulation study of the aerosol information content in OMI spectral reflectance measurements. *Atmospheric Chemistry and Physics*, **7**, 3115-3127.

Wang, J., S. Christopher, F. Brechtel, J. Kim, B. Schmid, J. Redemann, P. Russell, P. Quinn, and B. Holben, 2003: Geostationary satellite retrievals of aerosol optical thickness during ACE-Asia. *Journal of Geophysical Research*, **108**, 8657, 10.1029/2003JD003580.

Wang, S., Q. Wang, and G. Feingold, 2003: Turbulence, condensation and liquid water transport in numerically simulated nonprecipitating stratocumulus clouds. *Journal of the Atmospheric Sciences*, **60**, 262-278.

Warner, J., and S. Twomey, 1967: The production of cloud nuclei by cane fires and the effect on cloud droplet concentration. *Journal of the Atmospheric Sciences*, **24**, 704-706.

Warner, J., 1968: A reduction of rain associated with smoke from sugar-cane fires—An inadvertent weather modification. *Journal of Applied Meteorology*, **7**, 247-251.

Welton, E., K. Voss, P. Quinn, P. Flatau, K. Markowicz, J. Campbell, J. Spinhirne, H. Gordon, and J. Johnson, 2002: Measurements of aerosol vertical profiles and optical properties during INDOEX 1999 using micro-pulse lidars. *Journal of Geophysical Research*, **107**, 8019, doi:10.1029/2000JD000038.

Welton, E., J. Campbell, J. Spinhirne, and V. Scott, 2001: Global monitoring of clouds and aerosols using a network of micro-pulse lidar systems, in Lidar Remote Sensing for Industry and Environmental Monitoring, U. N. Singh, T. Itabe, N. Sugimoto, (eds.), *Proceedings of SPIE*, **4153**, 151-158.

Wen, G., A. Marshak, and R. Cahalan, 2006: Impact of 3D clouds on clear sky reflectance and aerosol retrieval in a biomass burning region of Brazil. *IEEE Geoscience and Remote Sensing Letters*, **3**, 169-172.

Wetzel, M. A. and Stowe, L. L.: Satellite-observed patterns in stratus microphysics, aerosol optical thickness, and shortwave radiative forcing. 1999: *Journal of Geophysical Research,*, **104**, 31287-31299.

Wielicki, B., B. Barkstrom, E. Harrison, R. Lee, G. Smith, and J. Cooper, 1996: Clouds and the Earth's radiant energy system (CERES): An Earth observing system experiment. *Bulletin of the American Meteorological Society*, **77**, 853-868.

Wild, M., H. Gilgen, A. Roesch, et al., 2005: From dimming to brightening: Decadal changes in solar radiation at Earth's surface. *Science*, **308**, 847-850.

Winker, D., R. Couch, and M. McCormick, 1996: An overview of LITE: NASA's Lidar In-Space Technology Experiment. *Proceedings of IEEE*, **84**(2), 164-180.

Winker, D., J. Pelon, and M. McCormick, 2003: The CALIPSO mission: spaceborne lidar for observation of aerosols and clouds. *Proceedings of SPIE*, **4893**, 1-11.

Xue, H., and G. Feingold, 2006: Large eddy simulations of tradewind cumuli: Investigation of aerosol indirect effects. *Journal of the Atmospheric Sciences*, **63**, 1605-1622.

Xue, H., G. Feingold, and B. Stevens, 2008: Aerosol effects on clouds, precipitation, and the organization of shallow cumulus convection. *Journal of the Atmospheric Sciences*, **65**, 392-406.

Yu, H., S. Liu, and R. Dickinson, 2002: Radiative effects of aerosols on the evolution of the atmospheric boundary layer. *Journal of Geophysical Research*, **107**, 4142, doi:10.1029/2001JD000754.

Yu, H., R. Dickinson, M. Chin, Y. Kaufman, B. Holben, I. Geogdzhayev, and M. Mishchenko, 2003: Annual cycle of global distributions of aerosol optical depth from integration of MODIS retrievals and GOCART model simulations. *Journal of Geophysical Research*, **108**, 4128, doi:10.1029/2002JD002717.

Yu, H., R. Dickinson, M. Chin, Y. Kaufman, M. Zhou, L. Zhou, Y. Tian, O. Dubovik, and B. Holben, 2004: The direct radiative effect of aerosols as determined from a combination of MODIS retrievals and GOCART simulations. *Journal of Geophysical Research,* **109**, D03206, doi:10.1029/2003JD003914.

Yu, H., Y. Kaufman, M. Chin, G. Feingold, L. Remer, T. Anderson, Y. Balkanski, N. Bellouin, O. Boucher, S. Christopher, P. DeCola, R. Kahn, D. Koch, N. Loeb, M. S. Reddy, M. Schulz, T. Takemura, and M. Zhou, 2006: A review of measurement-based assessments of aerosol direct radiative effect and forcing. *Atmospheric Chemistry and Physics,* **6**, 613-666.

Yu, H., R. Fu, R. Dickinson, Y. Zhang, M. Chen, and H. Wang, 2007: Interannual variability of smoke and warm cloud relationships in the Amazon as inferred from MODIS retrievals. *Remote Sensing of the Environment,* **111**, 435-449.

Yu, H., L.A. Remer, M. Chin, H. Bian, R. Kleidman, and T. Diehl, 2008: A satellite-based assessment of trans-Pacific transport of pollution aerosol. *Journal of Geophysical Research,* **113**, D14S12, doi:10.1029/2007JD009349.

Zhang, J., and S. Christopher, 2003: Longwave radiative forcing of Saharan dust aerosols estimated from MODIS, MISR, and CERES observations on Terra. *Geophysical Research Letters,* **30**, 2188, doi:10.1029/2003GL018479.

Zhang, J., S. Christopher, L. Remer, and Y. Kaufman, 2005a: Shortwave aerosol radiative forcing over cloud-free oceans from Terra. I: Angular models for aerosols. *Journal of Geophysical Research,* **110**, D10S23, doi:10.1029/2004JD005008.

Zhang, J., S. Christopher, L. Remer, and Y. Kaufman, 2005b: Shortwave aerosol radiative forcing over cloud-free oceans from Terra. II: Seasonal and global distributions. *Journal of Geophysical Research,* **110**, D10S24, doi:10.1029/2004JD005009.

Zhang, J., J. S. Reid, and B. N. Holben, 2005c: An analysis of potential cloud artifacts in MODIS over ocean aerosol optical thickness products. *Geophysical Research Letters,* **32**, L15803, doi:10.1029/2005GL023254.

Zhang, J., J.S. Reid, D.L. Westphal, N.L. Baker, and E.J. Hyer, 2008: A system for operational aerosol optical depth data assimilation over global oceans. *Journal of Geophysical Research,* **113**, doi:10.1029/2007JD009065.

Zhang, Q. et al., 2007: Ubiquity and dominance of oxygenated species in organic aerosols in anthropogenically-influenced Northern Hemisphere midlatitudes. *Geophysical Research Letters,* **34**, L13801, doi:10.1029/2007GL029979.

Zhang, X., F.W. Zwiers, and P.A. Stott, 2006: Multi-model multi-signal climate change detection at regional scale. *Journal of Climate,* **19**, 4294-4307.

Zhang, X., F.W. Zwiers, G.C. Hegerl, F.H. Lambert, N.P. Gillett, S. Solomon, P.A. Stott, T. Nozawa, 2006: Detection of human influence on twentieth-century precipitation trends. *Nature,* **448**, 461-465, doi:10.1038/nature06025.

Zhao, T. X.-P., I. Laszlo, W. Guo, A. Heidinger, C. Cao, A. Jelenak, D. Tarpley, and J. Sullivan, 2008a: Study of long-term trend in aerosol optical thickness observed from operational AVHRR satellite instrument. *Journal of Geophysical Research,* **113**, D07201, doi:10.1029/2007JD009061.

Zhao, T. X.-P., H. Yu, I. Laszlo, M. Chin, and W.C. Conant, 2008b: Derivation of component aerosol direct radiative forcing at the top of atmosphere for clear-sky oceans. *Journal of Quantitative Spectroscopy and Radiative Transfer,* **109**, 1162-1186.

Zhou, M., H. Yu, R. Dickinson, O. Dubovik, and B. Holben, 2005: A normalized description of the direct effect of key aerosol types on solar radiation as estimated from AERONET aerosols and MODIS albedos. *Journal of Geophysical Research,* **110**, D19202, doi:10.1029/2005JD005909.

Agricultural practices also affect air quality, such as leaving bare soil exposed to wind erosion, and burning agricultural waste. Photo taken from the NASA DC-8 aircraft during ARCTAS-CARB field experiment in June 2008 over California. Credit: Mian Chin, NASA.

Photography and Image Credits

Cover/Title Page/Table of Contents:

Image 1: Fire in the savanna grasslands of Kruger National Park, South Africa, during the international Southern African Fire-Atmosphere Research Initiative (SAFARI) Experiment, September 1992. Due to extensive and frequent burning of the savanna grass, Africa is the "fire center" of the world. Credit: Joel S. Levine, NASA.

Image 2: Urban pollution in Hong Kong, May 2007. The persistent pollution haze significantly reduces the visibility. Credit: Mian Chin, NASA.

Image 3: Dust storms of northwest Africa captured by Sea-viewing Wide Field-of-view Sensor (SeaWiFS) on February 28, 2000. Credit: SeaWiFS Project at NASA Goddard Space Flight Center.

Image 4: Breaking ocean waves – a source of sea salt aerosols. Credit: Mian Chin, NASA.

Image 5: Clouds at sunset. Clouds and aerosols scatter the sun's rays very effectively when the sun is low in the sky, creating the bright colors of sunrise and sunset. Credit: Mian Chin, NASA.

Image 6: Ship tracks appear when clouds are formed or modified by aerosols released in exhaust from ship smokestacks. Image from MODIS. Credit: NASA.

For other images in this report, please see the captions/credits located with each image.

Contact Information

Global Change Research Information Office
c/o Climate Change Science Program Office
1717 Pennsylvania Avenue, NW
Suite 250
Washington, DC 20006
202-223-6262 (voice)
202-223-3065 (fax)

The Climate Change Science Program
incorporates the U.S. Global Change Research
Program and the Climate Change Research
Initiative.

To obtain a copy of this document, place
an order at the Global Change Research
Information Office (GCRIO) web site:
http://www.gcrio.org/orders.

Climate Change Science Program and the Subcommittee on Global Change Research

William Brennan, Chair
Department of Commerce
National Oceanic and Atmospheric Administration
Director, Climate Change Science Program

Jack Kaye, Vice Chair
National Aeronautics and Space Administration

Allen Dearry
Department of Health and Human Services

Anna Palmisano
Department of Energy

Mary Glackin
National Oceanic and Atmospheric Administration

Patricia Gruber
Department of Defense

William Hohenstein
Department of Agriculture

Linda Lawson
Department of Transportation

Mark Myers
U.S. Geological Survey

Tim Killeen
National Science Foundation

Patrick Neale
Smithsonian Institution

Jacqueline Schafer
U.S. Agency for International Development

Joel Scheraga
Environmental Protection Agency

Harlan Watson
Department of State

EXECUTIVE OFFICE AND OTHER LIAISONS

Robert Marlay
Climate Change Technology Program

Katharine Gebbie
National Institute of Standards & Technology

Stuart Levenbach
Office of Management and Budget

Margaret McCalla
Office of the Federal Coordinator for Meteorology

Robert Rainey
Council on Environmental Quality

Daniel Walker
Office of Science and Technology Policy

www.ingramcontent.com/pod-product-compliance
Lightning Source LLC
Chambersburg PA
CBHW080641180526
45168CB00008B/3262